IEEE Radio and Antenna Days of the Indian Ocean (IEEE RADIO 2016)

IOP Conference Series: Materials Science and Engineering
Volume 198

Reunion Island, France
10 – 13 October 2016

Editor:

Vikass Monebhurrun

ISBN: 978-1-5108-4040-9
ISSN: 1757-8981

Printed from e-media with permission by:

Curran Associates, Inc.
57 Morehouse Lane
Red Hook, NY 12571

proceedings.com

Some format issues inherent in the e-media version may also appear in this print version.

Printed by Curran Associates, Inc. (2017)

Institute of Physics
Dirac House, Temple Back
Bristol BS1 6BE UK

Phone: 44 1 17 929 7481
Fax: 44 1 17 920 0979

techtracking@iop.org

Additional copies of this publication are available from:

Curran Associates, Inc.
57 Morehouse Lane
Red Hook, NY 12571 USA
Phone: 845-758-0400
Fax: 845-758-2633
Email: curran@proceedings.com
Web: www.proceedings.com

Table of contents

Volume 198

IEEE Radio and Antenna Days of the Indian Ocean
10–13 October 2016, Réunion Island, France

Accepted papers received: 2 May 2017
Published online: 11 May 2017

Preface

011001
OPEN ACCESS
EDITOR'S FOREWORD

011002
OPEN ACCESS
Editorial Board Members

011003
OPEN ACCESS
Peer review statement

Papers

012001
OPEN ACCESS
All-dielectric microwave devices for controlling the path of electromagnetic waves

J Yi, A de Lustrac, G-P Piau and S N Burokur.....1

012002
OPEN ACCESS
Electrical characterization of two-dimensional materials and their heterostructures

H Arora, T Schönherr and A Erbe.....5

012003
OPEN ACCESS
Scattering Control Using Nonlinear Smart Metasurface with Internal Feedback

D V Semenikhina and A I Semenikhin.....9

012004
OPEN ACCESS
A new assessment method of pHEMT models by comparing relative errors of drain current and its derivatives up to the third order

Josef Dobeš, Martin Grábner, Pavel Puričer, František Vejražka, Jan Míchal and Jakub Popp.....13

012005
OPEN ACCESS
Co-simulation of a complete rectenna with a circular slot loop antenna in CPW technology

Jérôme Rivière, Alexandre Douyère, Jonathan Cazour, Frédéric Alicalapa and Jean-Daniel Lan Sun Luk.....17

012006
OPEN ACCESS
Linear Vector Quantisation and Uniform Circular Arrays based decoupled two-dimensional angle of arrival estimation

Joseph D. Ndaw, Andre Faye and Amadou S. Maïga.....21

012007
OPEN ACCESS
Joint Channel and Phase Noise Estimation in MIMO-OFDM Systems

I M Ngebani, J M Chuma, I Zibani, E Matlotse and K Tsamaase.....25

012008
OPEN ACCESS
Impact of in-band interference on a wake-up radio system in wireless sensor networks

J M Lebreton, N M Murad and R Lorion.....29

012009
OPEN ACCESS
Arc Fault Detection & Localization by Electromagnetic-Acoustic Remote Sensing

C Vasile and C Ioana.....33

012010
OPEN ACCESS
Comparison of RF Photonics-Based Beamformers for Super-Wide Bandwidth Phased Array Antennas

M. Belkin, A. Sigov, Y. Tyschuk and V. Golovin.....37

012011
OPEN ACCESS
Characteristics of Vivaldi antenna located on a cylindrical surface

Y V Yukhanov, A V Gevorkyan and T Y Privalova.....41

012012
OPEN ACCESS
Radio Frequency Interference Detection using Machine Learning.

Olorato Mosiane, Nadeem Oozeer, Arun Aniyan and Bruce A. Bassett.....45

012013
OPEN ACCESS
Classifying bent radio galaxies from a mixture of point-like/extended images with Machine Learning.

David Bastien, Nadeem Oozeer and Radhakrishna Somanah.....49

012014
OPEN ACCESS
MIGHTEE: The MeerKAT International GHz Tiered Extragalactic Exploration

A Russ Taylor and Matt Jarvis.....53

IOP Conf. Series: Materials Science and Engineering **198** (2017) 011001 doi:10.1088/1757-899X/198/1/011001

EDITOR'S FOREWORD

Dr. Philippe Jean Pierre, President of the Regional Committee for Innovation of Reunion Island, officially opened the fourth edition of the Radio and Antenna Days of the Indian Ocean (IEEE RADIO 2016) that was held at Hotel Le Récif, Saint-Gilles Les Bains, Reunion Island from 10th to 13th October 2016. The conference was organized by the Radio Society (Mauritius) and it was jointly sponsored by the IEEE Antennas and Propagation Society (APS) and Université de La Réunion. In addition, IEEE RADIO 2016 was technically co-sponsored by Union Radio Scientifique Internationale (URSI), IEEE Region 8 and IEEE France Section. The local organizing committee consisted of researchers from CentraleSupélec, France, Université de La Réunion and University of Mauritius.

IEEE RADIO 2016 is the fourth of a series of conferences, covering the fields of antennas and propagation, organized in the Indian Ocean region. The idea behind the organization of the RADIO international conference originated at the start of this decade following discussions with engineers and researchers from countries of the Indian Ocean region as well as from other parts of the world. It was unanimously acknowledged that there was an urgent need to initiate a platform that would provide engineers and researchers of the Indian Ocean region opportunities to share knowledge in the field of radio frequency engineering with their international peers. IEEE RADIO 2016 gathered participants from over twenty countries coming from different regions of the world including Africa, Asia, Australia, Europe, Middle-East and North America, and we can confidently say that we have succeeded in our goal.

The conference featured fourteen oral sessions on state-of-the-art research themes including antenna design, bio-electromagnetics, computational electromagnetics, devices and circuits, metamaterials and applications, radio astronomy, remote sensing and wireless sensor networks. Invited talks were delivered by Prof. Tapan Sarkar from Syracuse University, USA, Prof. Qing Liu from Duke University, USA and Prof. Monaï Kraikriksh from King Mongkut's Institute of Technology, Ladkrabang, Thailand. A special session, organized by Dr. Eric Mokole, Vice-Chair of IEEE APS Special Interest Group on Humanitarian Technology (SIGHT), was dedicated to humanitarian activities and projects in the Indian Ocean.

All accepted papers that were effectively presented during the conference were submitted to the IEEE Xplore Digital Library. And, in line with the previous editions, a call to extended contributions for publication as a volume in the IoP Conference Series: Materials Science and Engineering (MSE) was made. All of the twenty full papers that were submitted were then peer-reviewed, revised according to the reviewers' comments and ultimately fourteen papers were accepted for publication in this volume.

This fourth edition of the IoP Conference Series: MSE dedicated to selected papers presented at IEEE RADIO 2016 is representative of the scientific content and quality of the RADIO international series of conferences.

Editor
Vikass Monebhurrun

IOP Conf. Series: Materials Science and Engineering **198** (2017) 011002 doi:10.1088/1757-899X/198/1/011002

Editorial Board Members

IOP Conf. Series: Materials Science and Engineering **198** (2017) 011003 doi:10.1088/1757-899X/198/1/011003

Peer review statement

All papers published in this volume of *IOP Conference Series: Materials Science and Engineering* have been peer reviewed through processes administered by the proceedings Editors. Reviews were conducted by expert referees to the professional and scientific standards expected of a proceedings journal published by IOP Publishing.

IOP Conf. Series: Materials Science and Engineering **198** (2017) 012001 doi:10.1088/1757-899X/198/1/012001

All-dielectric microwave devices for controlling the path of electromagnetic waves

J Yi[1], A de Lustrac[2,3], G-P Piau[4] and S N Burokur[5,6]
[1] State Key Laboratory of Integrated Services Networks, Xidian University, 710071 Xi'an, Shaanxi, China
[2] Centre de Nanosciences et de Nanotechnologies, CNRS, Univ Paris Sud, Université Paris-Saclay, 91405 Orsay cedex, France
[3] Université Paris Nanterre, 92410 Ville d'Avray, France
[4] AIRBUS Group Innovations, 92150 Suresnes, France
[5] LEME, EA 4416, Université Paris Ouest, 92410 Ville d'Avray, France

E-mail: sburokur@u-paris10.fr

Abstract. All-dielectric devices are designed using Quasi-Conformal Transformation Optics (QCTO) concept and fabricated by additive manufacturing for the control of wave propagation. Three lenses are studied; the first one is used to compensate for the curvature of a non-planar antenna array, the second one to steer an electromagnetic beam to a desired direction and the last one to taper an electromagnetic field between two sections of different dimensions.

1. Introduction

Transformation Optics (TO) is a powerful tool that enables to control electromagnetic (EM) fields in unprecedented and unbelievable ways through the use of judiciously engineered materials with parameters that vary spatially [1, 2]. The TO formalism is based on the invariance of Maxwell's equations over space-time coordinate transformations. Such flexibility in controlling EM waves appears to be convenient in the design of innovative devices with performances or special properties difficult to achieve from classical methods and has therefore inspired considerable research interests in wave propagation. The electromagnetic cloak is one of the most striking devices realized with TO concept [3].

In order to avoid anisotropy in material parameters that arises during coordinate transformation, an equivalence between coordinate transformation and spatial deformation by using Laplace's equation to determine the deformation of coordinate grids during the transformation has been proposed [4]. So the transformed material parameters are determined by the solutions of the Laplace's equation under proper boundary conditions. Such mapping known as quasi-conformal transformation optics (QCTO) has been proposed to simplify the electromagnetic distributions in the new transformed coordinate space. QCTO helps to minimize the anisotropy of the constitutive materials, giving the possibility to implement all-dielectric materials in the transformed medium. As a result, nearly-isotropic gradient index (GRIN) materials with broad frequency bandwidth and low losses can be employed, opening the way to broadband devices.

[6] To whom any correspondence should be addressed.

IOP Conf. Series: Materials Science and Engineering **198** (2017) 012001 doi:10.1088/1757-899X/198/1/012001

In this work, QCTO is applied to design microwaves devices for controlling the path of electromagnetic waves. Two broadband lenses; one capable of restoring in-phase emissions from an array of radiators fixed on a non-planar ground plane and another one capable of steering the radiated beam of an antenna to an off-normal direction and a waveguide taper are designed, fabricated and experimentally validated in this work.

2. QCTO formulations

The design of a QCTO-based device is associated to the calculation of the 2D Laplace's equation so as to evaluate the coordinate grids deformation during the transformation from the initial virtual space filled with vacuum in the coordinate system (x, y, z) to the real final physical space filled with transformed medium in the coordinate system (x', y', z') [4]. The method then implies that the calculation of the electromagnetic material parameters of the transformation medium is equivalent to the computation of spatial deformation field governed by the 2D Laplace's equation with proper boundary conditions, whatever the shape of the device.

The conformal mapping in the physical space is obtained by solving Laplace's equation with predefined boundary conditions. Let's suppose that the coordinate transformation between the physical and virtual spaces is $x = f(x', y')$, $y = g(x', y')$. The mathematical equivalence of the transformation is expressed by the Jacobian matrix $J = \partial(x, y)/\partial(x', y')$. By solving Laplace's equations in the virtual space with respect to specific boundary conditions

$$\frac{\partial^2 x}{\partial x'^2} + \frac{\partial^2 x}{\partial y'^2} = 0, \quad \frac{\partial^2 y}{\partial x'^2} + \frac{\partial^2 y}{\partial y'^2} = 0 \tag{1}$$

the Jacobian matrix J of the mapping can be obtained.

The physical space performs an inverse function of the virtual space. Thus the Jacobian matrix of this inverse transformation from (x', y') to (x, y) is represented by J^{-1}. The properties of the intermediate medium are calculated such that the constitutive tensors have only diagonal components. It can then be found that the mapping satisfies the vector form of Laplace's equation. This leads to the simple material parameters given as

$$\varepsilon = \mu = \begin{bmatrix} 1 & 0 & 0 \\ 0 & 1 & 0 \\ 0 & 0 & |J|^{-1} \end{bmatrix} \tag{2}$$

For fields' equivalence at the outer boundaries with the virtual space for which the transformation is carried out, Neumann and Dirichlet sliding boundary conditions are applied to fix the geometry of the transformed domain and to maintain orthogonality on the boundary. The determination of the mapping is performed using COMSOL Multiphysics Partial Differential Equation (PDE) solver. In all the designed lenses, the transformation deals with a two-dimensional (2D) model with incident transverse electric (TE) polarized wave. In such a configuration, the electric field has only a component directed along the z direction. The material properties of the lenses are then simplified and possess only dielectric properties $\varepsilon = \dfrac{\varepsilon_r}{\det(J^{-1})}$.

3. QCTO-based devices

First, we study the enhancement of directivity through the use of a QCTO-based lens conformed on a non-planar array of radiating elements that follows a cylindrical shape. The lens is capable of restoring in-phase emissions from a conformal array of antennas so as to obtain performances similar to that of a linear one. The design of the conformal lens-antenna is presented in Fig. 1(a). Calculations done in [5-6] have shown that the effective permittivity varies from 1 to 2.76 in the lens.

The second lens is designed to steer an antenna's radiated beam to an off-normal direction. The lens-antenna structure is shown in Fig. 2(a). The permittivity values range from 1 to 6.4 in the designed lens [7-8]. A linear array of 4 patch antennas is used as primary source.

The third device is a coupling device capable of tapering the electromagnetic fields between two sections of different dimensions (Fig. 3(a)). The electromagnetic waves exiting from a large section or aperture are properly guided through the taper to a smaller one. The permittivity (ε_{zz}) distribution ranges from 1 to 5.02 for the calculated taper.

4. Implementations and experimental validations

In this work, we favor the realization of the permittivity values by using a dielectric with air holes. We consider 3D printing additive fabrication technique. The polymer used in the printing process presents a relative permittivity $\varepsilon_d = 2.8$ [9].

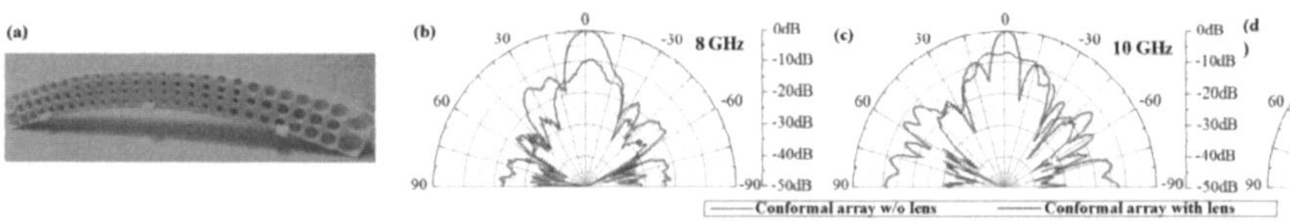

Figure 1. (a) Conformal lens for in-phase emission restoring. (b)-(d) Measured performances in the 8 GHz – 12 GHz frequency range.

Taking in consideration the minimum and maximum value of hole diameter that can be realized, we fix the variation of the effective permittivity from 1 to 2.8 in the discrete model of the devices. Such variation is possible by varying the diameter of the holes.

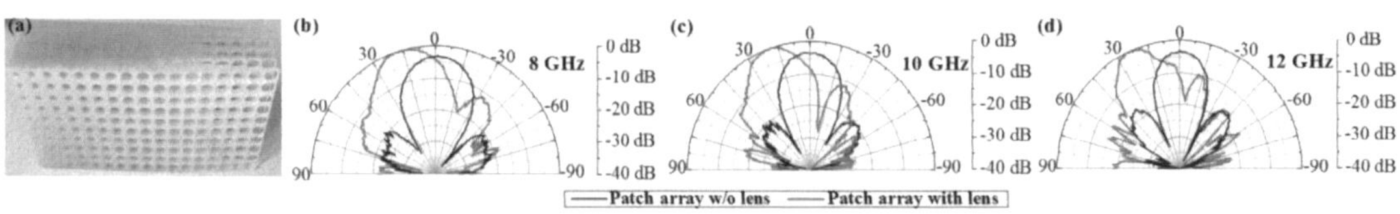

Figure 2. (a) Beam steering lens for off-normal radiation. (b)-(d) Measured performances in the 8 GHz – 12 GHz frequency range.

The performances measured from the transformed devices illustrate clearly the control of the path of electromagnetic waves. The all-dielectric devices present the advantage of having broadband characteristics. Theoretically, the achieved bandwidth can be very broad since we are making use of all-dielectric non resonant materials. However, in physically fabricated prototypes, the bandwidth will definitely depend on the size of the devices and also on the operating frequency. At low frequencies, the size of the device must be large enough since wavelength is large and at high frequencies, we are limited by small wavelength. Therefore at high frequencies the unit cells (holes in dielectric) used to tailor the material parameters must be engineered with respect to the wavelength so as to be consistent with the effective medium theory. Therefore, a trade-off has to be necessarily made between the size of the lens and the desired bandwidth so that the transformed medium can be considered homogeneous in such frequency range and so that the dimensions of the device are large enough at the lowest frequency of operation.

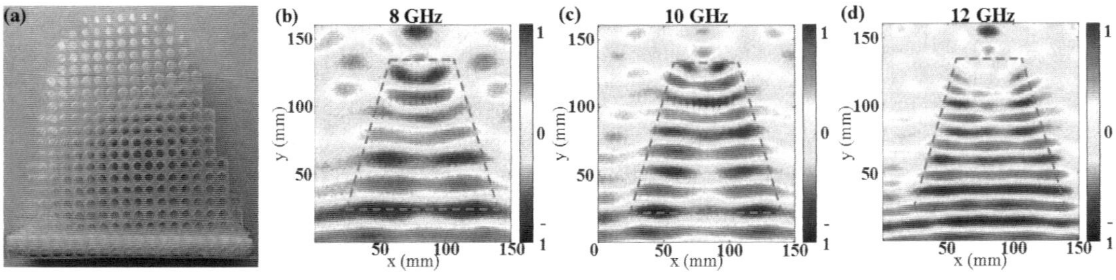

Figure 3. (a) Taper to control the flow of electromagnetic waves between two waveguides of different cross-sections. (b)-(d) Measured performances in the 8 GHz – 12 GHz frequency range.

5. Conclusion

To summarize, we have presented the design and experimental realization of compact all-dielectric microwave devices operating on a wide frequency range. The devices are designed through the transformation optics concept in order to control the direction of EM wave propagation. Furthermore, the design concept has been combined with additive fabrication technology for the realization of the devices. The concept can be readily applied at any frequency with available fabrication techniques.

References

[1] Leonhardt U 2006 *Science* **312** 1777
[2] Pendry J B, Schurig D and Smith D R 2006 *Science* **312** 1780
[3] Schurig D, Mock J J, Justice B J, Cummer S A, Pendry J B, Starr A F and Smith D R 2006 *Science* **314** 977
[4] Hu J, Zhou X and Hu G 2009 *Opt. Express* **17** 1308
[5] Yi J, Burokur S N, Piau G-P and de Lustrac A 2015 *Appl. Phys. Lett.* **107** 024101
[6] Yi J, Burokur S N, Piau G-P and de Lustrac A 2016 *J. Opt.* **18** 044010
[7] Yi J, Burokur S N and de Lustrac A 2015 *Opt. Express* **23** 12942
[8] Yi J, Burokur S N, Piau G-P and de Lustrac A 2016 *Sci. Rep.* **6** 18819
[9] Yi J, Piau G-P, de Lustrac A and Burokur S N 2016 *Sci. Rep.* **6** 30661

IOP Conf. Series: Materials Science and Engineering **198** (2017) 012002 doi:10.1088/1757-899X/198/1/012002

Electrical characterization of two-dimensional materials and their heterostructures

H Arora[1,2], T Schönherr[1] and A Erbe[1]

[1]Helmholtz-Zentrum Dresden-Rossendorf, Bautzner Landstrasse 400, 01328 Dresden, Germany
[2]Technische Universität Dresden, 01062 Dresden, Germany

Email: h.arora@hzdr.de

Abstract. Two-dimensional (2D) materials have gained enormous attention in recent years owing to their huge potential in future electronics and optics. On the one hand, conventional 2D materials like graphene, MoS_2, h-BN are being intensively studied, on the other hand, search for novel 2D materials is at a rapid pace. In this study, we have investigated electrical properties of 2D nanosheets of ultrathin Indium Selenide (InSe), a member of the III-VI chalcogenides' family. The InSe layers were prepared via micromechanical cleavage of its bulk crystal and were integrated into a field-effect transistor (FET) device as the transport channel. On characterizing the InSe-based FETs, InSe showed n-type conductance with the mobility of 2.1×10^{-4} $cm^2V^{-1}s^{-1}$.

1. Introduction

The electronic industry is teeming with new devices. An ever-increasing challenge to beat Moore's law has led to the development of devices which are as small and compact as possible. The transistors today are already reaching nanometers (nm) of channel length [1]. However, as the devices are approaching sub-10 nm dimensions, it's becoming increasingly difficult to follow Moore's law. Silicon (Si), the dominating material in all electronic devices so far, is predicted to fail the sub-10 nm technology transistors because of severe short channel effects [2]. With this respect, two-dimensional (2D) materials have gained enormous attention in recent years owing to their intrinsic properties which compensate for the drawbacks of Si at such small dimensions [3-5]. Due to quantum confinement, they exhibit properties different from their bulk phases and thus, offer an opportunity to carry forward further miniaturization of electronic devices.

2D materials are theoretically known since decades [6-7], however, their first experimental proof came only in 2004, when Graphene was isolated from its bulk form Graphite [8]. Since then the field of 2D materials is booming, leading to more than 600 2D materials currently known [9-10]. The family of 2D materials consists of conductors, insulators and semiconductors, which are being intensively researched for their applications in the fields of electronics, optics and sensing [9-11].

Semiconducting 2D materials having a finite band gap, excellent transport properties and mechanical flexibility are of particular interest for the electronic industry. Transition metal-dichalcogenides (TMDCs) like MoS_2, WSe_2 are one of the most prominent classes of 2D materials, showing excellent semiconducting properties and thus, being researched world-wide [12-14]. Another class of semiconducting 2D materials rapidly gaining interest is the family of III-VI chalcogenides. These are layered materials with the formula–

MX (M=Ga, In; X=S, Se, Te), arranged in X-M-M-X tetra-atomic thick layers [5]

Their thermal stability and absence of dangling bonds make them highly suitable for nano-electronic and photonic applications [15]. Though many theoretical investigations have been done for these materials, very limited experimental knowledge is available regarding their performance in nano-electronics [9, 16-17].

Within the III-VI chalcogenide family, we have focused our study primarily on Indium Selenide (InSe). One of the major reasons is its relatively high stability as compared to other members of the family, which gave us the chance to characterize it over a length of time. Moreover, InSe has lighter electron effective mass ($m^* = 0.143\ m_o$) and shows higher mobility (μ) of ~ $10^3\ cm^2V^{-1}s^{-1}$ as compared to MoS_2 ($m^* = 0.45\ m_o$, $\mu = 50–200\ cm^2V^{-1}s^{-1}$) [15] and thus, can be used for fast, high performance electronics where MoS_2 has been proved undesirable. In this study, we have investigated electrical properties of atomically thin layers of InSe in order to understand the underlying transport phenomena. InSe consists of vertically stacked 2D sheets of Se-In-In-Se and has a honeycomb lattice as shown in figure 1. The atoms in the same layer are held by covalent bonds, whereas inter-layer bonding is via Van-der-Waals forces. In bulk form, InSe has a direct band gap of 1.3 eV at room temperature [5].

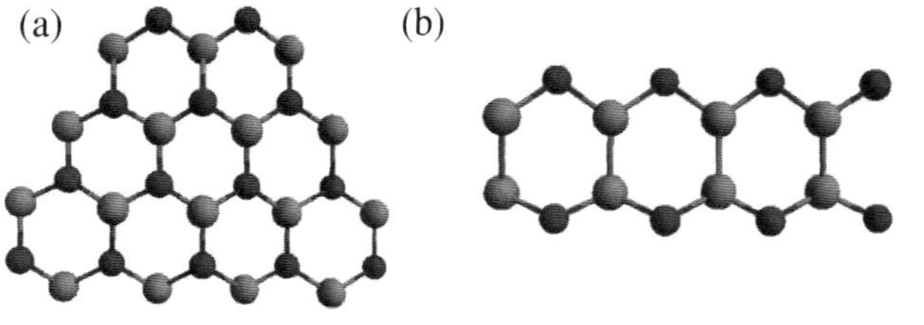

Figure 1. (a) Honeycomb structure of 2D InSe, (b) layers of Se-In-In-Se stacked vertically to give the structure of monolayer InSe. Color scheme: Indium in brown, Selenium atoms in blue.

2. Experimental Section

In order to understand the electrical properties of 2D InSe layers, we fabricated InSe-based field-effect transistors (FETs) in back gate configuration, where the transport channel was atomically thin layer of InSe.

2.1. Micromechanical cleavage of InSe ultrathin layers

Atomically thin layers of InSe were obtained via micromechanical cleavage of its bulk crystal. The InSe bulk crystal (2H-InSe) used for this study was bought commercially from 2D Semiconductors with a purity of 99.995%. The bulk crystal was repeatedly peeled off with the help of a Scotch-tape to obtain few and single layers of InSe, which were deposited directly on highly p-doped Silicon (Si) substrates with 283 nm of SiO_2 layer serving as back gate for the fabricated devices. The identification of single and few layers of InSe was done using the techniques of optical microscopy and atomic force microscopy (AFM) which have been discussed in later sections.

2.2. Device fabrication

On top of the deposited InSe layers, metal electrodes (source and drain) were defined using high-precision electron-beam lithography (EBL). The electrodes were fabricated by depositing a Ti (5 nm)/Au (100 nm) layer by electron-beam evaporation, followed by standard lift-off process in acetone.

2.3. Device characterization

The optical images of the InSe nanoflakes were taken with an optical microscope (OLYMPUS BX51). The thickness and roughness of the flakes were determined by tapping mode AFM in air using a Bruker Multimode 8 scanning probe microscope. The electrical measurements of the fabricated devices were carried out in back gate configuration in a semiautomatic Süss Microtec Probe System PA200 combined with a semiconductor characterization system (4200-SCS, Keithley Instruments). To maintain same environmental conditions during the electrical measurements of the devices, the temperature of the wafer chuck of the probe system was maintained at 25.0±0.5 °C by using a

IOP Conf. Series: Materials Science and Engineering **198** (2017) 012002 doi:10.1088/1757-899X/198/1/012002

precision temperature control system. In addition to this, an air dehumidifier system was also used in order to maintain low humidity inside the probe system down to a dew point of -40 °C.

3. Results and Discussion

Figure 2(a) shows InSe layers deposited on Si/SiO_2 (283 nm) substrate via standard scotch tape technique as discussed above. In this method, different thicknesses ranging from single to bulk layers are deposited. However, one of the simplest and fastest techniques to distinguish between different layer thicknesses is optical imaging. Different layer thicknesses of InSe produce different color contrasts with respect to the interference of the underlying SiO_2 layer on exposing them to the visible light. In figure 2(a) the change in color of the flakes with changing thicknesses can be easily seen. Thinner layers give a low contrast, which improves with increasing thickness or number of layers. Using AFM we confirmed the thicknesses of the selected layers as shown in figure 2(b).

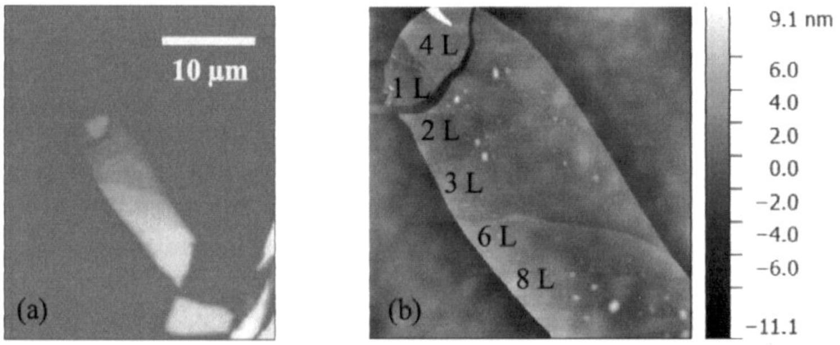

Figure 2. (a) Optical image and (b) AFM of InSe flakes on Si/SiO_2 substrate; the number of layers– 1 L, 2 L... are labelled as measured under AFM.

In our study, we have focused on electrical properties of ultrathin layers of InSe i.e. single or few layer structures. The optical image of one of our fabricated devices from InSe bilayer structure is shown in figure 3(a) and schematic of bottom-gate FET device on ultrathin InSe layer in figure 3(b).

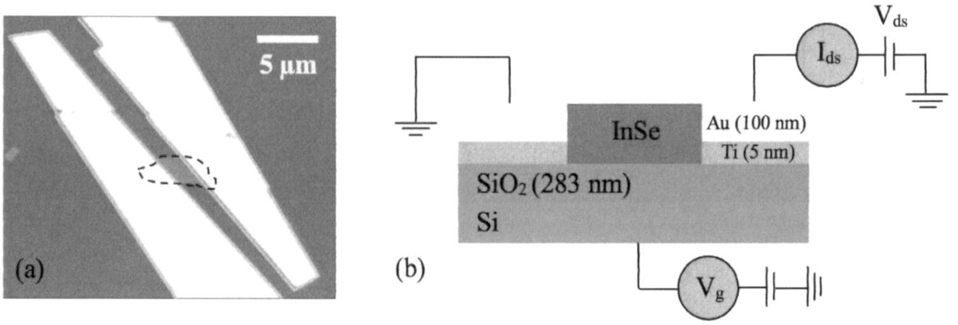

Figure 3. (a) InSe layer (dashed) 2.3 nm thick (as determined by AFM) contacted with Ti/Au metal electrodes, (b) schematic of bottom-gate FET, InSe serving as the transport channel.

The channel length i.e. distance between the contact pads for this device is approx. 2 µm. We measured the electrical characteristics of the FET by applying a drain-source voltage (V_{ds}) and a back gate voltage (V_{bg}) as shown in the schematic in figure 3(b). The measured output characteristic (Figure 4(a)) shows that Au metal did not form ohmic contacts to the InSe layer. The non-linearity in the negative bias and no current flow in the positive bias shows formation of a Schottky contact. The transfer characteristics (Figure 4(b)) shows semiconducting n-type behavior with mobility (µ) of 2.1×10^{-4} $cm^2V^{-1}s^{-1}$, estimated using the expression of field-effect mobility:

$$\mu = [dI_{ds}/dV_{bg}] \times [L/W] \times [1/C_{ox}V_{ds}] \text{ at } V_{ds} = 20V \quad (1)$$

where I_{ds} is the drain-source current, V_{bg} is the back gate voltage, L and W are length and width of the channel respectively, C_{ox} is the capacitance per unit area of the SiO_2 layer and V_{ds} is the drain-source voltage. It should be noted that our values for the field-effect mobility are underestimated due to the formation of Schottky contacts and indulgence of very high contact resistance in our two-probe measurements.

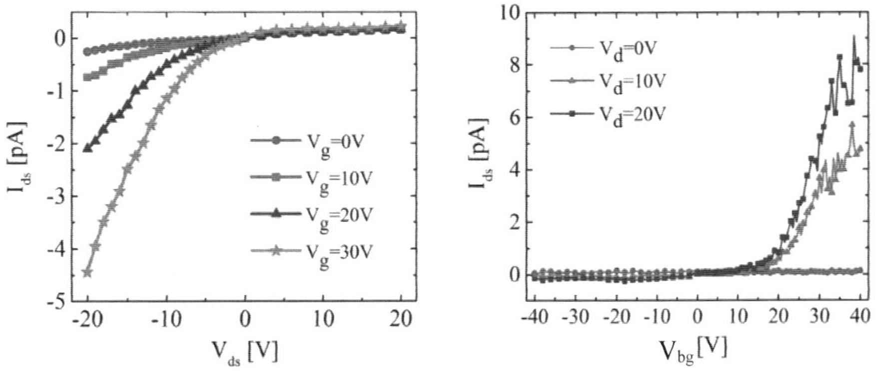

Figure 4. I–V curves for InSe bilayer FET (a) I_{ds}–V_{ds} curve for V_{bg} sweep from 0 to 30 V, (b) I_{ds}–V_{bg} curve for V_{ds} sweep from -1 to 1V.

4. Conclusion

In this work, we have investigated ultrathin layers of InSe, a non-conventional 2D material which still lacks in-depth experimental studies. The layers showed good electrical properties w.r.t. transistor performance and an intrinsic n-type behavior. However, the mobility obtained was very low as compared to theoretical estimations, which was attributed to high contact resistance due to formation of Schottky contact between Au and InSe layer or to the degradation of InSe surface in the ambience.

Acknowledgement

We thank Mr. Holger Lange for assistance with the probe system and Mr. Vinayakrishna Joshi for AFM measurements. This work was kindly supported by the Initiative and Networking Fund of the Helmholtz Association of German Research Centers through the International Helmholtz Research School for Nanoelectronic Networks, IHRS NANONET (VH-KO-606).

References

[1] "Intel Discloses Newest Microarchitecture and 14 Nanometer Manufacturing Process Technicak Details". Intel. August 11 2014.
[2] M Lundstrom 2003 *Science* **299** 210–11.
[3] W Cao, J Kang, D Sarkar, W Liu and K Banerjee 2015 *IEEE Trans.Electron.Dev.* **62** 3459–69.
[4] G Fiori, F Bonaccorso, G Iannaccone, T Palacios, D Neumaier, A Seabaugh, S K Banerjee and L Colombo 2014 *Nature Nanotechnology* **9** 768–79.
[5] K Xu, L Yin, Y Huang, T A Shifa, J Chu, F Wang, R Cheng, Z Wanga and J He 2016 *Nanoscale* **8** 16802.
[6] J A Wilson and A D Yoffe 1969 *Advances in Physics* **18** 193–335.
[7] J V McCanny and R B Murray 1977 *J. Phys. C: Solid State Phys.* **10** 1211–22.
[8] K S Novoselov, A K Geim, S V Morozov, D Jiang, Y Zhang, S V Dubonos, I V Grigorieva and A A Firsov 2004 *Science* **306** 666–69.
[9] P Miro, M Audiffred and T Heine 2014 *Chem. Soc. Rev.* **43** 6537–54.
[10] K S Novoselov, A Mishchenko, A Carvalho, A H Castro Neto 2016 *Science* **353** aac9439-1.
[11] X Duan, C Wang, A Pan, R Yu and X Duan 2015 *Chem. Soc. Rev.* doi: 10.1039/c5cs00507h
[12] B Radisavljevic, A Radenovic, J Brivio, V Giacometti and A Kis 2011 *Nature Nanotechnology* **6** 147–150.
[13] H Liu, A T Neal and P D Ye 2012 *ACS Nano* **6** 8563–69.
[14] D Sarkar, X Xie, W Liu, W Cao, J Kang, Y Gong, S Kraemer, P M Ajayan and K Banerjee 2015 *Nature* **526** 91–95.
[15] W Feng, W Zheng, W Cao and P Hu 2014 *Adv. Mater.* **26** 6587–93.
[16] W Wei, Y Dai, C Niu, X Li, Y Mad and B Huanga 2015 *J. Mater. Chem. C* **3** 11548.
[17] Y Ma, Y Dai, M Guo, L Yu and B Huang 2013 *Phys.Chem. Chem. Phys.* **15** 7098.

IOP Conf. Series: Materials Science and Engineering **198** (2017) 012003 doi:10.1088/1757-899X/198/1/012003

Scattering Control Using Nonlinear Smart Metasurface with Internal Feedback

D V Semenikhina[1] **and A I Semenikhin**[1]

[1]Southern Federal University, Taganrog, Russia

E-mail: d_semenikhina@mail.ru

Abstract. The ideology of creation of a nonlinear smart metasurface with internal feedback for the adaptive control by spectral composition of scattered field is offered. The metasurface contains a lattice of strip elements with nonlinear loads-sensors. They are included in a circuit of internal feedback for the adaptive control of scattered field. Numerically it is shown that maximal levels of the second harmonic in the spectrum of scattered far field correspond to maximum of voltage rectified on metasurface. Experimentally the prototype of the plane smart covering on the basis of the metasurface in the form of strip lattice with controlled nonlinear loads-sensors is investigated for an idea confirmation.

1. Introduction

Nonlinear metasurfaces (NMSs) on the basis of controlled nonlinear loads (NLs) are applied in different microwave, terahertz and optical devices [1-3]. The NMSs can potentially contribute to developing a new type of an applications of smart artificial surface [2]. Designing of smart NMSs requires the organization of adaptive control of nonlinear scattering. However, generally in a control system of scattering there is no trustworthy information about characteristics of far field.

Irradiation by means of the harmonic field of nonlinear metasurfaces with controlled NLs leads to appearance of the side reradiation at frequencies of harmonics and combinative spectral components of the main signal. Levels of the reradiated fields at frequencies of harmonics are functionally connected to the NL parameters, in particular, with a voltage rectified on NLs with the square voltage-current characteristic (VCC). Thus, NLs with the controlled VCC parameters allow to control the spatial-frequency spectrum of scattered field [4].

In this paper the creation of a nonlinear smart metasurface with internal feedback for the adaptive control by spectral composition of a scattered field is considered. The metasurface contains a lattice of strip elements with nonlinear loads-sensors included in a circuit of internal feedback.

2. Analysis of theoretical results

In paper [4] the solution of a problem of a scattering of a plane monochromatic wave by the infinite periodic lattice of the strip elements (the left part of figure 1) having nonlinear loads (NLs) was resulted using method of nonlinear integral equations. Unit cell of investigated NMS on the substrate with ground plane is shown in the right part of figure 1. It is built on the base of the nonlinear lattice. The metasurface contains a lattice of strip elements with nonlinear loads. The VCC of NL was set as

$$i^e = \sum_{\nu=0}^{P} \left(a_\nu u(t)^\nu + b_\nu du^\nu(t)/dt \right),$$

IOP Conf. Series: Materials Science and Engineering **198** (2017) 012003 doi:10.1088/1757-899X/198/1/012003

where i, u are current through load and voltage on load terminal pair; a_V, b_V are the coefficients determined by electrophysical properties of load; t is time.

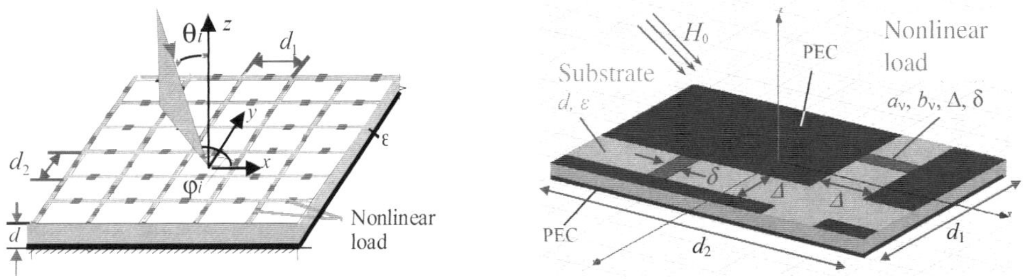

Figure 1. Statement of the problem of scattering of incident plane wave (with amplitude H_0 and angles incident of θ_i, φ_i) by the infinite periodic lattice of the strip elements with NLs (left) and unit cell of nonlinear metasurface on the substrate with ground plane (PEC): $d_1 = d_2 = 5$ mm, $\Delta = \delta = 1$ mm, $d = 0.3$ mm, $\varepsilon = 2.74$ (right).

The lattice of strip elements was located on the dielectric substrate of permeability ε and thickness d above the ground.

Stages of solution included [5]: obtaining of nonlinear boundary conditions regarding harmonics of electric and magnetic surface currents $J_{nx}^{e1,2}, J_{ny}^{m1,2}$ on loads; obtaining of the nonlinear integral equation system; obtaining of nonlinear algebraic equation system; numerical solution of the nonlinear algebraic equation system regarding harmonics of magnetic surface currents on z=0-plane; defining of harmonics of scattering field or other scattering characteristics. Dependences of the constant component of the voltage (induced by incident wave) on inputs of NL [4] as well as dependences of lattice reflection coefficients from parameters of loads and the incident plane electromagnetic wave are investigated. Value of rectified (in the load) voltage U_0 is defined by values of zero-order harmonics of magnetic surface currents on NLs (they found in result of the numerical decision of system of nonlinear integral equations [4]). Reflection coefficients $\left|R_n^{jk}\right|$ show the attitude of amplitudes of (j,k)-th Floquet's harmonics of fields, reflected from the latticeat frequencies of n-th harmonics, to amplitude of the incident field at fundamental frequency f_1 (in case of n=1).

Let us analyze the dependences of the rectified (on the load) voltage versus the bias voltage applied to the load and of the distance from the scatterer to the source of the probing signal. We will consider that charts of rectified voltages versus bias voltage are similar to dependences of reflection coefficients versus coefficients of loads' polynomial VCC. Charts of rectified voltage versus distance are similar to dependences of reflection coefficients versus amplitude of the irradiating wave. In figure 2 rectified voltage versus coefficient a_2 of square term of the VCC is shown. The sizes of the lattice are set so that at these frequencies they belong the single-mode area of periodicity. Calculations are given for the passive lattices having loads with the positive linear conductivity a_1. Rectified voltage U_0 with growth of a_2 increases quadratically at first, and then almost linearly. In case of some coefficient $a_2 = a_{2\max}$ the voltage reaches the maximum. The a_{2max} is defined by linear admittance of the load as well as absolute and relative geometrical sizes of the load (in comparison with the unit cell sizes). In figure 3 reflection coefficients of zero-order Floquet's mode $\left|R_n^{00}\right|$ at fundamental frequency (n=1), the second (n=2) and third (n=3) frequency harmonics versus coefficient a_2 of square term of the V C are shown. In the case of passive loads the reflection coefficient of zero-order Floquet's mode on the second frequency harmonic also reaches the maximum in the case of $a_2 = a_{2max}$.

Means, in case of this value of a_{2max}, the greatest redistribution of energy of incident wave at harmonic frequencies is carried out; the efficiency of conversion of wave energy to rectified voltage will be maximum. In the case of increase in values $a_2 > a_{2\max}$ values of $\left|R_2^{00}\right|$ and $U_0(a_2)$ decrease. It can be explained by a changing of a shape of load VCC from close to characteristic of diode detector (in case of $a_2 \le a_{2max}$) on almost parabolic (in case of $a_2 >> a_{2max}$).

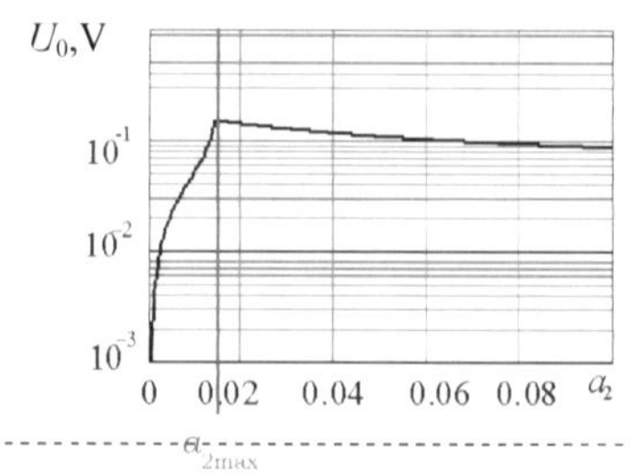

Figure 2. Rectified voltage U_0 versus coefficient a_2 of square term of the VCC.

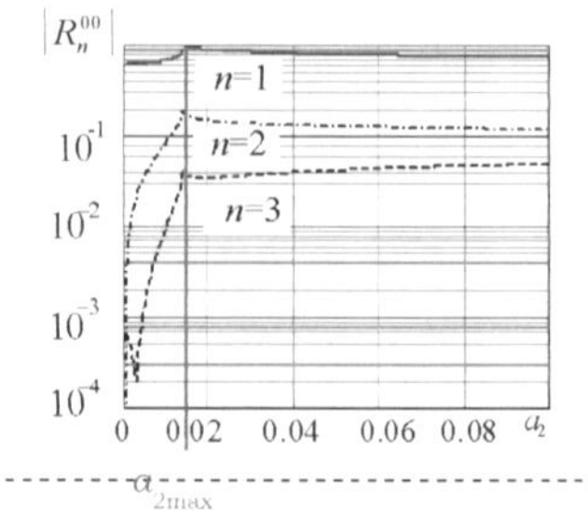

Figure 3. Reflection coefficients versus coefficient a_2 of square term of the VCC.

Nature of dependence of rectified voltage U_0 versus amplitude H_0 of incident wave (figure 4) is also defined by a ratio of VCC coefficients of NLs (i.e. a type of VCC). When VCC about origin of coordinates has the type close to the characteristic of the diode detector (curve 1* in inset of figure 4), for some H_0 the value of rectified voltage reaches a maximum, and then decreases (in some range of increase of H_0). In case of other VCC types of loads ("cubic" and "parabolic", see curves 2*, 3* in inset of figure 4), diagrams have the sections of square growth in case of $H_0 < 0.2$ A/m and slower growth in case of $H_0 > 0.2$ A/m.

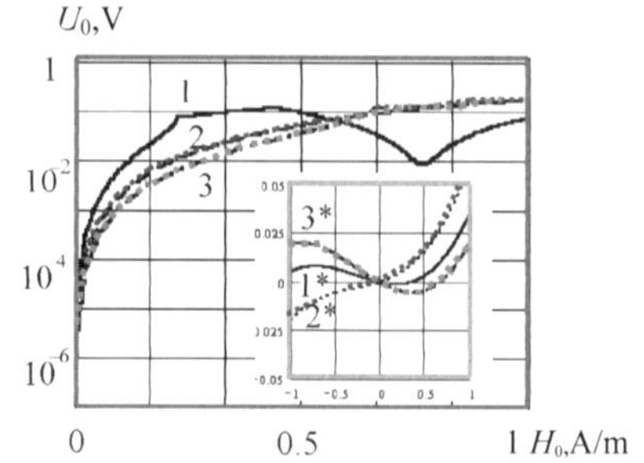

Figure 4. Rectified voltage versus of incident wave amplitude for θ_i=0.

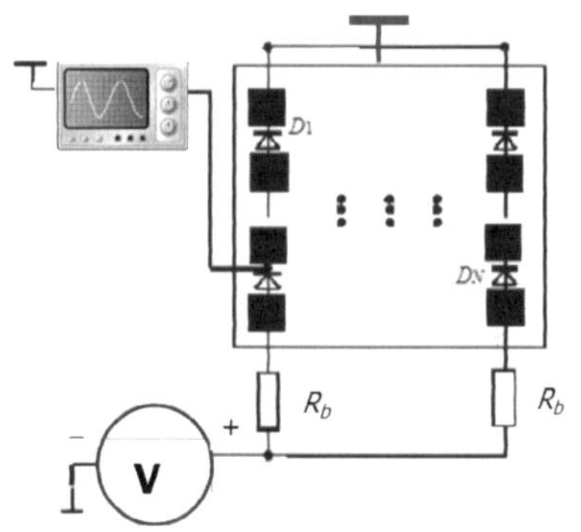

Figure 5. Equivalent scheme circuit of the prototype.

3. The analysis of the experimental characteristics of the prototype

The prototype of the plane smart covering on the basis of NMS with internal natural feedback (figure 5) represents the finite lattice of square strip elements. Between their edges the diodes are placed in one of the directions. On the diodes the bias voltage is served. From the diodes at the same time with bias voltage the rectified voltage is removed. For achievement of necessary bias voltage, diodes are switched on sequentially on columns through balance resistance, and columns of diodes are paralleled. Rectified voltage was measured on one diode or on a column of diodes. The photograph of a prototype of the microstrip lattice with nonlinear loads is shown in figure 6.

In figure 7 the dependence of the rectified voltage on one diode, measured across the integrating capacity, versus bias voltage is shown. It is visible that there is the optimum of bias voltage given on one diode. At the same time the maximum of rectified voltage is watched that is unambiguous connected to a maximum of far field on the second harmonic n = 2 (see figure 2).

IOP Conf. Series: Materials Science and Engineering **198** (2017) 012003 doi:10.1088/1757-899X/198/1/012003

Also the rectified voltage on a column of diodes is measured versus amplitude of the irradiating field E_0. It is visible that results of experiment (figure 8) and calculation (figure 4) match qualitatively.

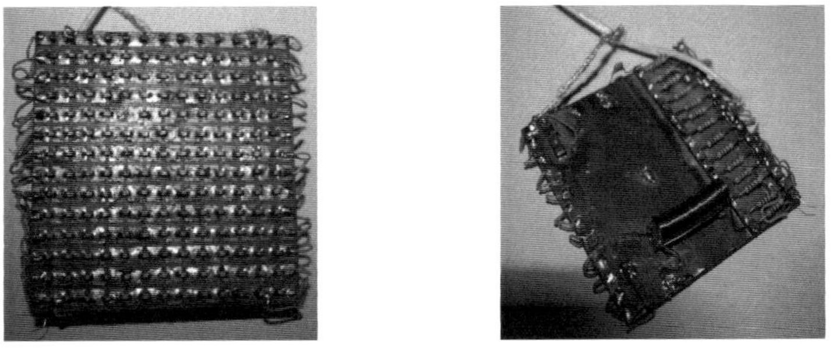

Figure 6. The prototype of the microstrip lattice with nonlinear loads: front side (left) and back side (right).

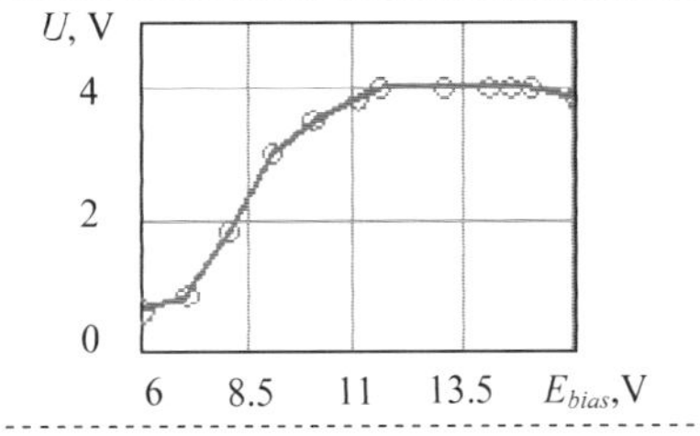

Figure 7. Rectified voltage versus bias voltage; f_1=9.4 GHz.

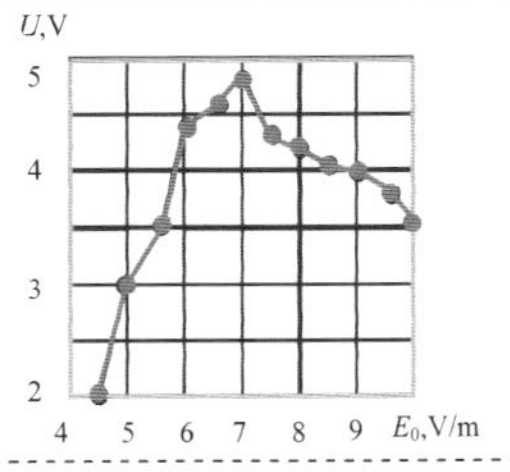

Figure 8. Rectified voltage versus amplitude of the irradiating field E_0; f_1=9.4 GHz.

4. Conclusion

The obtained results have allowed to offer the ideology of creation of a nonlinear smart metasurface with internal feedback for the adaptive (to the changing exitation conditions) control by spectral composition of far field. Information about the rectified voltage and tuning of bias voltage toward maximum of the rectified voltage provide the greatest transfer of energy at the frequency of the second harmonic and enrichment of a spectrum of the reradiated signal.

Acknowledgement

This research was financially supported by Russian Science Foundation (project No. 16-19-10537).

References

[1] Shadrivov I V, Lapine M and Kivshar Y S (Eds) 2015 *Nonlinear, Tunable and Active Metamaterials* (Springer)

[2] Sievenpiper D, Quarfoth R, Wakatsuchi H and Luo Zhangjie 2015 *9th European Conference on Antennas and Propag. (EuCAP)* pp 1-2

[3] Wakatsuchi H, Kim S, Rushton J and Sievenpiper D 2013 Applied Physics Letters, **102**, 214103-4

[4] Semenikhina D V and Semenikhin A I 2011 *Int. Conf. on Electromagnetics in Advanced Applications (ICEAA)* (Torino: Italy)

[5] Semenikhina D V and Dekalo I E 1998 *14 Internat. Wroclaw Symp. on Electromag. Compat.* (Wroclaw: Poland)

IOP Conf. Series: Materials Science and Engineering **198** (2017) 012004 doi:10.1088/1757-899X/198/1/012004

A new assessment method of pHEMT models by comparing relative errors of drain current and its derivatives up to the third order

Josef Dobeš, Martin Grábner, Pavel Puričer, František Vejražka, Jan Míchal, and Jakub Popp

Department of Radioelectronics, Faculty of Electrical Engineering, Czech Technical University in Prague

E-mail: dobes@fel.cvut.cz, mgrabner@cmi.cz, {puricep, vejrazka, michal, poppjaku}@fel.cvut.cz

Abstract. Nowadays, there exist relatively precise pHEMT models available for computer-aided design, and they are frequently compared to each other. However, such comparisons are mostly based on absolute errors of drain-current equations and their derivatives. In the paper, a novel method is suggested based on relative root-mean-square errors of both drain current and its derivatives up to the third order. Moreover, the relative errors are subsequently relativized to the best model in each category to further clarify obtained accuracies of both drain current and its derivatives. Furthermore, one our older and two newly suggested models are also included in comparison with the traditionally precise Ahmed, TOM-2 and Materka ones. The assessment is performed using measured characteristics of a pHEMT operating up to 110 GHz. Finally, a usability of the proposed models including the higher-order derivatives is illustrated using s-parameters analysis and measurement at more operating points as well as computation and measurement of IP_3 points of a low-noise amplifier of a multi-constellation satellite navigation receiver with ATF-54143 pHEMT.

1. Introduction

Selecting and identifying a suitable pHEMT model for radio electronics is not a simple problem. Probably the most challenging tasks are simulations of the (critical) mixing $2f_1 - f_2$ and $2f_2 - f_1$ products because such an analysis needs credible 3rd-order derivatives of the model functions. Generally, traditional simpler models like Curtice, Statz, and TOM [1] are used as well as more accurate ones as TOM-2 [2], Ahmed [3], Parker-Skellern [4], or Materka [5]. Besides that, we have also suggested some modified and novel models [6]. All these models including ours were often compared as shown in [3], e.g. However, most of the comparisons are based on absolute values, especially those of higher-order derivatives. Therefore, we suggest a novel way of comparison based on ratios of relative errors of the models. Furthermore, we propose and demonstrate additional (indirect) methods for the assessment of the model accuracy.

2. Unabbreviated definition of proposed TANH and XEXP models

An abbreviated definition of the suggested TANH and XEXP models has been shown in [6]. The complete equations are given as

$$i_{DS} = \begin{cases} g_{\min} v_{DS} & \text{for } v_{GS} - (V_{T0} - \gamma v_{DS}) \le 0 \text{ and} \\ & N_G + N_D v_{DS} = 0, \\ g_{\min} v_{DS} + \beta V_G^{+}(1 + \lambda v_{DS}) \tanh(\alpha v_{DS}) & \text{otherwise,} \end{cases} \tag{1}$$

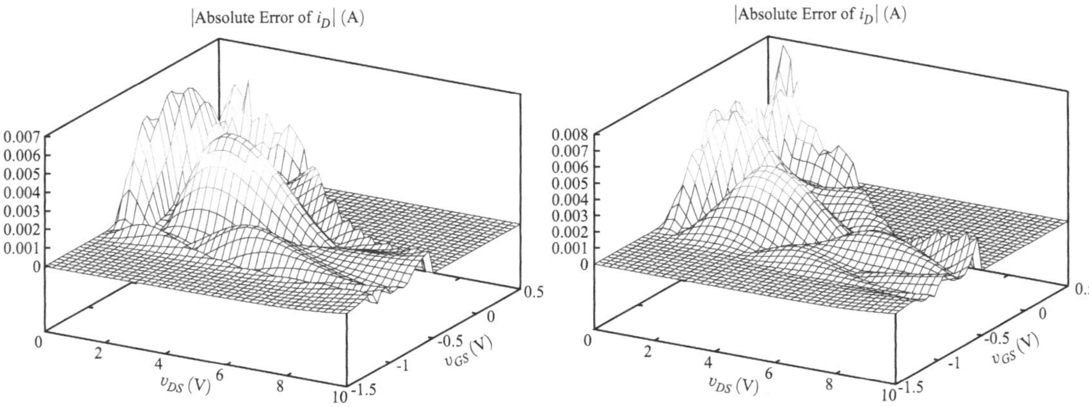

Figure 1. Comparison of absolute errors of the Ahmed (left) and XEXP (right) models used for 110 GHz 0.25 µm pHEMT [7]. Although the XEXP model has a higher peak in the two-dimensional error graphs, it has lesser errors in the central part of characteristics, where pHEMT mostly works. (And it is important.)

where two *principally* new formulae were suggested for the V_G^+ factor to characterize a transconductance dependence (inclusion of the V_G^+ factor creates a new additional layer in the hierarchy of model equations):

$$V_G^+ = \begin{cases} V_{TS} \ln\left[\cosh\left(\frac{V_G}{V_{TS}}\right)\right] & \text{for the TANH model,} \\ V_{TS}\left[\mathrm{e} - \left(1 + \frac{V_G}{V_{TS}}\right)\exp\left(1 - \frac{V_G}{V_{TS}}\right)\right] & \text{for the XEXP model.} \end{cases} \tag{2}$$

In this concept, the transconductance is estimated as $\tanh(\frac{V_G}{V_{TS}})$ for the TANH model, and by integrating the term $\frac{V_G}{V_{TS}}\exp(1 - \frac{V_G}{V_{TS}})$ for the XEXP model. V_{TS} (voltage of transconductance slope) is a new model parameter and for the V_G factor, formulae were slightly simplified (V'_{ST}) from those in the TOM-2 model:

$$V_G = V'_{ST} \ln\left[\mathrm{e}^{\frac{\upsilon_{GS} - (V_{TO} - \gamma \upsilon_{DS})}{V'_{ST}}} + 1\right], \quad V'_{ST} = 2(N_G + N_D \upsilon_{DS})\frac{kT}{q}. \tag{3}$$

All the other DC model parameters are defined in [1]. The formulae for model capacitances are defined in Table 2. They have been a bit modified in comparison with [2] (a saturation has been included where υ_{new} is used); however, except for $V_{\max}$, there are no new model parameters comparing with the definition [2].

3. Comparison of relative root-mean-square errors

First, we have calculated DC characteristics and their absolute errors for all the tested models with typical results for the drain current and its derivatives as shown in Figures 1 and 2. However, unlike in the majority of other papers, we have also computed *relative* errors not only for the drain current as usual, but also for all its derivatives up to the third order. Furthermore, we have relativized all these errors to the one of the model with the lowest error. All the models which had this error more than ten times greater than that of the best model in a category of the derivatives were excluded (i.e., Curtice, Statz, and TOM). The results of this comparison are shown in Table 1, where the error of the best model in each category is transformed to unit level and hence denoted by $\boxed{1}$. This favors the Ahmed, TANH, and XEXP models.

4. Checking the model by s-parameter measurement

We have also analyzed and measured pHEMT s-parameters at many operating points, which could also be considered an important test of the model precision. The results have been satisfactory, a typical example is shown in Figure 4, where s_{11}, $s_{12} \times 5$, $s_{21}/5$, and s_{22} plots are shown at $\upsilon_{GS} = -0.2$ V and $\upsilon_{DS} = 5$ V.

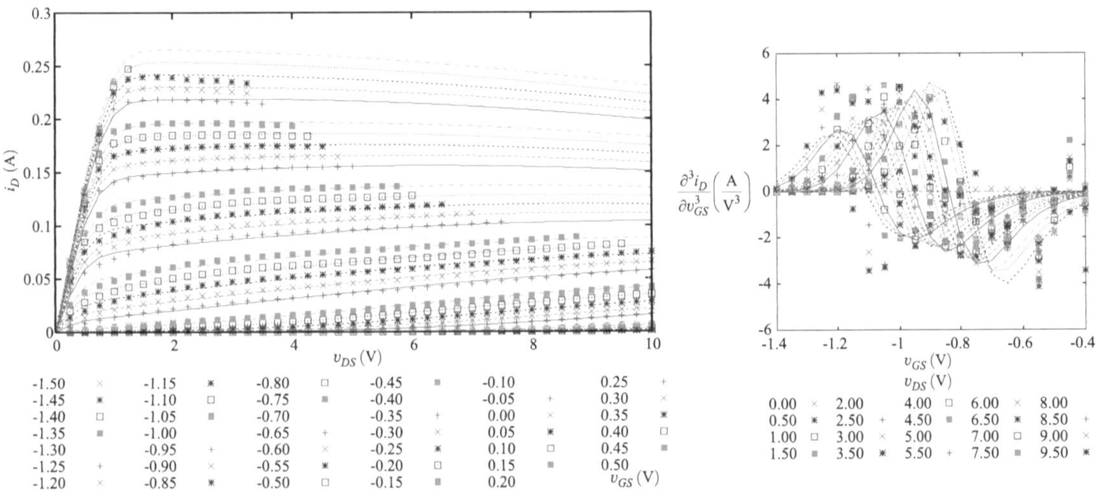

Figure 2. Modeling of the pHEMT drain current and its 3rd-order derivative $\frac{\partial^3 i_D}{\partial v_{GS}^3}$ by the TANH model (this derivative is important for accurate estimation of the $2f_2 - f_1$ mixer product). Note that the measured (points) and simulated (lines) values of the 3rd-order derivatives naturally differ for greater values of v_{DS}.

Table 1. Assessing the *relative* root-mean-square errors of the drain current and its derivatives.

Model	i_D	$\frac{\partial i_D}{\partial v_{DS}}$	$\frac{\partial^2 i_D}{\partial v_{DS}^2}$	$\frac{\partial^3 i_D}{\partial v_{DS}^3}$	$\frac{\partial i_D}{\partial v_{GS}}$	$\frac{\partial^2 i_D}{\partial v_{GS}^2}$	$\frac{\partial^3 i_D}{\partial v_{GS}^3}$
Ahmed	4.2	2.51	1	1	1.24	4.69	1
Dobes	4.49	6.49	1.29	1.13	1.6	7.6	3.35
TOM2	1.66	4.38	1.99	1.85	2.08	1.9	7.5
Materka	4.21	3.5	1.52	1.2	1.45	5.54	3.13
TANH	1	2.51	1.12	1.15	1.11	1	1.02
XEXP	1.01	1	1.27	1.2	1	4.46	1

5. Checking the model by third-order intercept point

A natural possibility of checking the 3rd-order derivatives consists in comparing computed and measured IP_3 points of an RF circuit with pHEMT at more operating points and frequencies. For this purpose, we have used a low-noise amplifier of a multiconstellation [8] receiver for satellite navigation with ATF-54143 [9] in Figure 3. The couples of parameters in the red indicate values before and after the multiobjective optimization [10]. Typical obtained computed and measured IP_3 points are presented in Figure 5, which shows quite acceptable agreement between the results of the circuit simulation and measurement.

The difference between the computed and measured IP_3 was (0.75, 3.8) dB for the model extracted on the basis of the relative errors, and (1.25, 1.33) dB when the greater absolute errors got greater weights.

6. Conclusions

A full definition of two suggested novel pHEMT models has been described in the paper. The models put emphasis on precise defining of higher-order derivatives, which is especially important for RF circuits. Moreover, the new method of relative error comparisons gave a meaningful score of the models, and the s-parameters analysis and measurement as well as IP_3 simulation and computation showed a viability.

Acknowledgments: This paper has been supported by the Technology Agency of the Czech Republic under the grant No. TE01020186, and by the grant SGS16/164/OHK3/2T/13 of the Czech Technical University in Prague.

Table 2. Capacitance model equations.

$$c_{GS} = \frac{K_1 K_2 C_{GS0}}{\sqrt{1 - \frac{\min(v_{\mathrm{new}}, V_{\max})}{\phi_0}}} + K_3 C_{GD0}$$

$$c_{GD} = \frac{K_1 K_3 C_{GS0}}{\sqrt{1 - \frac{\min(v_{\mathrm{new}}, V_{\max})}{\phi_0}}} + K_2 C_{GD0}$$

$$K_1 = \frac{1}{2}\left[1 + \frac{v_{\mathrm{ef}} - V_{T0}}{\sqrt{(v_{\mathrm{ef}} - V_{T0})^2 + V_\delta^2}}\right]$$

$$K_2 = \frac{1}{2}\left[1 + \frac{v_{GS} - v_{GD}}{\sqrt{(v_{GS} - v_{GD})^2 + (1/\alpha)^2}}\right]$$

$$K_3 = \frac{1}{2}\left[1 - \frac{v_{GS} - v_{GD}}{\sqrt{(v_{GS} - v_{GD})^2 + (1/\alpha)^2}}\right]$$

$$v_{\mathrm{ef}} = \frac{1}{2}\left[v_{GS} + v_{GD} + \sqrt{(v_{GS} - v_{GD})^2 + \left(\frac{1}{\alpha}\right)^2}\right]$$

$$v_{\mathrm{new}} = \frac{1}{2}\left[v_{\mathrm{ef}} + V_{T0} + \sqrt{(v_{\mathrm{ef}} - V_{T0})^2 + V_\delta^2}\right]$$

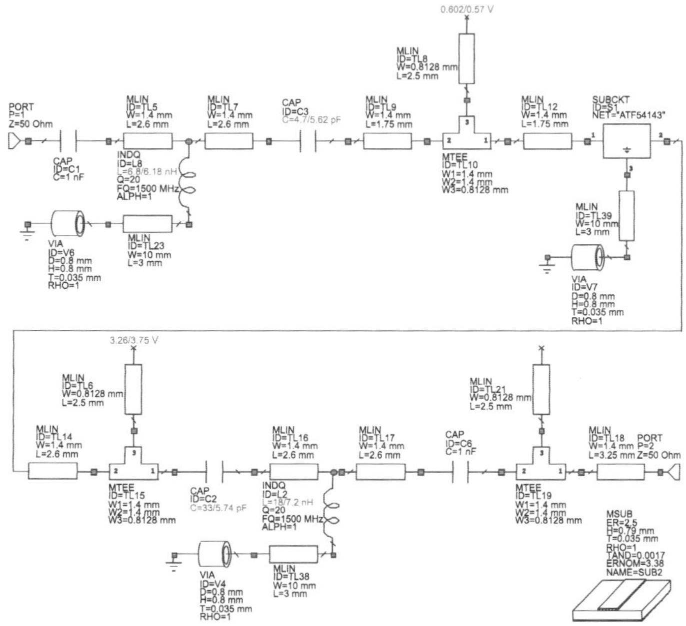

Figure 3. LNA amplifier for the multi-constellation receiver used for checking the calculated and measured IP_3 after optimization.

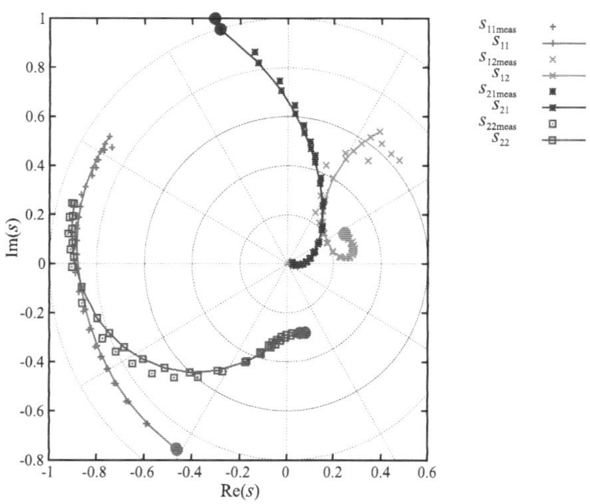

Figure 4. Calculated (lines) and measured (points) frequency dependencies of s-parameters in complex plane for frequencies from 4 (bold dots) to 90 GHz.

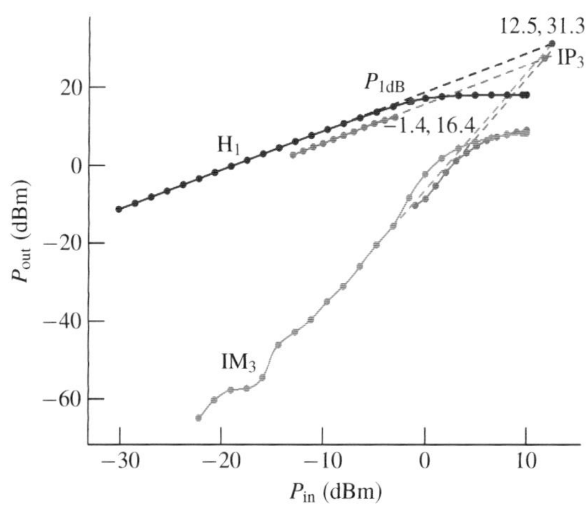

Figure 5. Computation (blue & green) and measurement (gray) of IP_3 for the frequencies f_1 and f_2 1575 and 1576 MHz (at equal levels), respectively.

References

[1] Cadence Design Systems 2003 *PSpice A/D Reference Guide* 10th ed

[2] Smith D H 1995 TOM-2: An improved model for GaAs MESFETs Tech. rep. TriQuint Semiconductor, Beaverton, OR

[3] Memon N M, Ahmed M M and Rehman F 2007 *Solid-State Electronics* **51**(3) 511–516

[4] Parker A E and Skellern D J 1997 *IEEE Transactions on Microwave Theory and Techniques* **45**(9) 1563–1571

[5] Materka A and Kacprzak T 1985 *IEEE Transactions on Microwave Theory and Techniques* **33**(2) 129–135

[6] Dobeš J and Grábner M 2011 *Proc. of IEEE Compound Semiconductor IC Symp.* (Waikoloa Village, Hawaii) pp 189–192

[7] Raffo A, Santarelli A, Traverso P A, Vannini G, Palomba F, Scappaviva F, Pagani M and Filicori F 2005 *IEEE Transactions on Microwave Theory and Techniques* **53**(11) 3449–3459

[8] Matějka Š 2016 *Radioengineering* **25**(2) 390–398

[9] June 2012 ATF-54143 low noise enhancement mode pseudomorphic HEMT data sheet Tech. Rep. AV02-0488EN Avago

[10] Dobeš J, Míchal J, Popp J, Černý D, Grábner M, Vejražka F, Kákona J and Matějka Š 2015 *Radioengineering* **24** 670–680

IOP Conf. Series: Materials Science and Engineering **198** (2017) 012005 doi:10.1088/1757-899X/198/1/012005

Co-simulation of a complete rectenna with a circular slot loop antenna in CPW technology

Jérôme Rivière, Alexandre Douyère, Jonathan Cazour, Frédéric Alicalapa and Jean-Daniel Lan Sun Luk

University of La Reunion, Laboratory of Electronic, Energetic and Process St Denis, Reunion Island

jerome.riviere@univ-reunion.fr

Abstract. This study starts with the design of a planar and compact CPW antenna fabricated on Arlon AD1000 substrate, ε_r=10.35. The antenna is a coplanar waveguide (CPW) fed circular slot loop antenna matched to the standard impedance 50 Ω by two stubs. The goal is to implement this antenna with a CPW RF/DC rectifier to build an optimized low power level rectenna. The rectenna design is restricted to allow easy and fast fabrication of an array with a high reproducibility. The full rectenna is simulated and achieves 10% effciency at -20 dBm.

1. Introduction

Lastly, with the growth of wireless communications and the development of smart cities, energy harvesting has received considerable attention [1]. Wireless communication developments tend to provide reliable and constant wireless energy. Those electromagnetic waves convey information but they are also energy carriers. The majority of the communicative devices spread their signals to all directions in space and so energy is wasted in the environment. The rectenna depict in figure 1 converts electromagnetic waves to DC voltage and can harvest this energy to power small devices. Robustness and repeatability in rectenna should be as perfect as possible to limit differences across samples of the same model. Moreover, many applications need power characteristics that require the use of an array of antenna. But, as the complexity of antenna and frequency number increase, phase shifting and complex matching impedance circuitry are needed. Another solution relies on rectenna arrays, which combine several rectenna DC voltages to power a device. This second solution requires reliable and accurate rectenna conception.

This work is dedicated to rectenna arrays solution. In order to suppress vias in rectenna conception we use CPW (Figure 2) [2]. In this technology, the conductive line width W and the ground plane are etched on the same plane, separated by a slot of width S on each side of the conductive line.

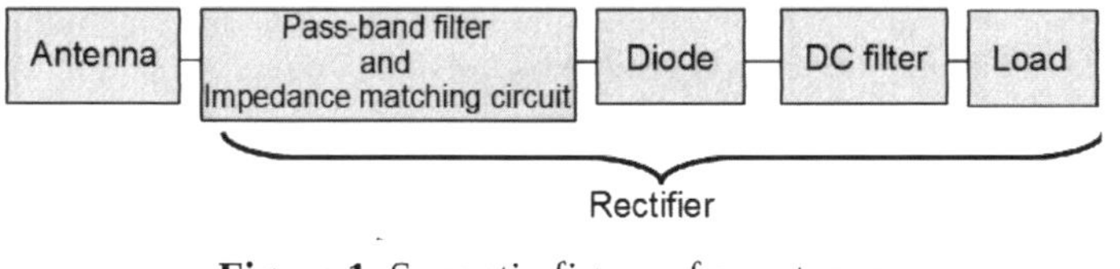

Figure 1. Synoptic figure of a rectenna

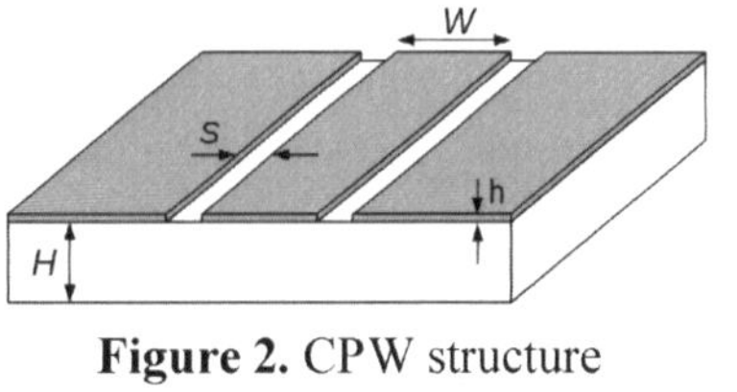

Figure 2. CPW structure

The topic of this paper focuses on the design and conception of a planar and compact antenna to be used with a RF/DC rectifier in CPW technology. This antenna is also simple and fast to realize on a CNC machine as few processing steps are required. It consists of a patch surrounded by a ground conductor on the same plane of the substrate, without background metallization to avoid the connection of a transition to the ungrounded CPW rectifier.

This paper begins with the conception and measurement of the CPW-fed circular slot antenna. In this section we have developed the steps leading to the final antenna design. The next section summarizes the conception of the CPW rectifier. Finally, the antenna is simulated with the rectifier and the efficiency is evaluated using an accurate diode model optimized for 2.45 GHz.

2. Antenna Design and measurement

2.1. Design

The antenna for this rectenna conception is a modified CPW fed circular slot loop antenna (or coplanar patch antenna). The study starts with a circular monopole antenna because it is one of the simplest patterns to study [3]. Only two parameters were optimized: patch radius (***R***) and distance (***d***) between patch and the ground. To facilitate fabrication process of the rectenna, we set the slot widths in all our designs to the 0.787 mm milling tool we use to etch the rectenna. The simulated initial antenna (Figure 3.a) shows a large bandwidth of 2.7 GHz, with a gain ***G*** of 1.4 dBi and directivity of 3.16 dBi. The return loss is of -28 dB at 2.45 GHz, and the total dimensions of the patch are 45*60 mm^2.

When plotting the current distribution of the patch at 2.45 GHz (Figure 3,a), we can notice that the current propagates along the edge of the circular patch and is "null" at the top of the circular monopole pattern. Hence, shielding the patch with a ground plane compels the magnetic field to propagate along the newly created slot (Figure 3.b); thus increasing the current at the edges while maintaining similar resonant frequency.

Shielding the circular pattern has indeed increase the current propagating along the edge of the antenna. This modification also modifies the frequency of operation and impedance. To correct the frequency shift, a diminution of the patch radius is sufficient. To match the 50 Ω impedance required from the rectifier we use stubs impedance matching technique because it offers advantages in the antenna design, such as size reduction. Hence, two stubs are inserted inside the radiating element to match the antenna impedance at 50 Ω (Figure 4.a). This solution gives a more compact antenna with a shorter feed line. Plus, the width of the feeding line remains equal to the width W of the rectifier conductive line.

It can be noted that the radius of the final circular patch is closed to the radius of the monopole antenna with only 0,5 mm difference. The use of stub in this case, because of the nature of the CPW feeding line can also be seen as a movement of the feed-point closer to the center of the circular patch. Thus, the antenna size is reduced to 45.2*50.57 mm^2 with a gain of 2.34 dBi, directivity of 3.87 dBi and peak current of 29.8 A/m (Figure 4.b).

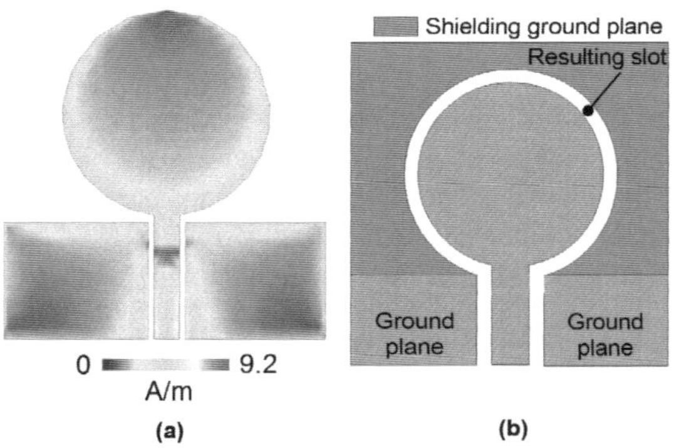

Figure 3. (a) Initial antenna current propagation and (b) Monopole antenna transformation to slot loop antenna

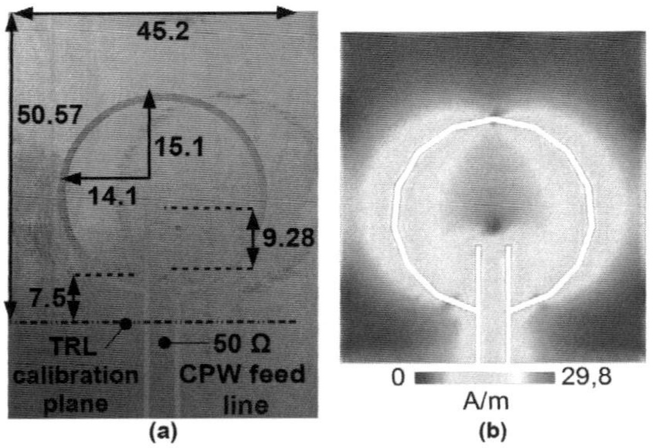

Figure 4. (a) CPW-fed circular slot loop antenna with dimensions in mm and (b) current propagation of the proposed antenna

2.2. Measurement

The proposed antenna is realized and measured (Figure 5). Measurement presents a bandwidth ***B*** of 250 MHz and return loss of -23.7 dB at 2.45 GHz. The simulation and measurement show good agreement with a small frequency drift ***ΔF*** of 10 MHz. To measure this antenna, we use TRL calibration technique. This allows us to take into account part of the ground plane, which will be added by the rectifier on the antenna performances [4].

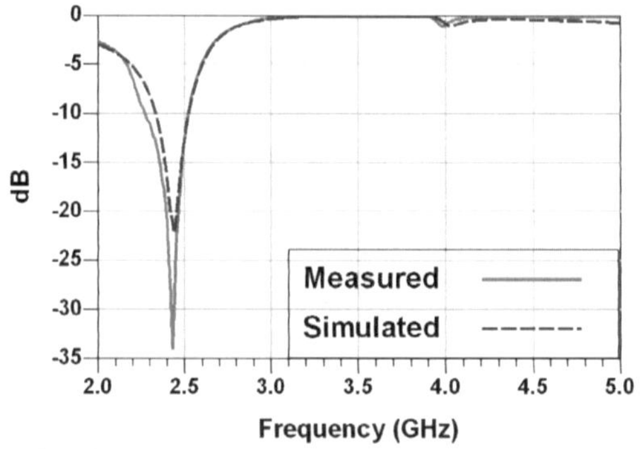

Figure 5. Reflection coefficient of the simulated and realized antenna

3. Rectenna simulation

3.1. Rectenna conception summary

In order to evaluate the complete rectenna efficiency, we use a co-simulation under Schematics environment on Advanced Design System 2013. Each element of the rectenna has been optimized to ease and speed up realization time, but has also been modified to reduce the rectenna dimensions. Those aspects of this work should be further addressed in a different paper when dealing with the global rectenna conception. For the purpose of this paper we will only focus on the co-simulated rectenna efficiency.

In our previous work [5], the HSMS7630 diode model used was composed of a generic diode component filled with SPICE model from the data sheet. We then upgraded this model to take into account the effect of the diode surrounding such as CPW line, soldering and parasitic effects of the diode body. But this model turned out to be ineffective, because it did not properly modeled the DC behaviour of the diode.

To fix this issue, we use a different model (Figure 6) in which we first model the DC behaviour of the diode with a current generator $\boldsymbol{G_c}$ driven by the voltage $\mathbf{V_{ad}}$. Afterward, modifications were made to model the diode frequency behaviour through the fitting of the model generated S parameters and measured ones. The effectiveness of this model has been primarily verified with the comparison between simulated and measured impedance of a sub-part of the rectenna made with the CPW DC filter in [7] and the diode.

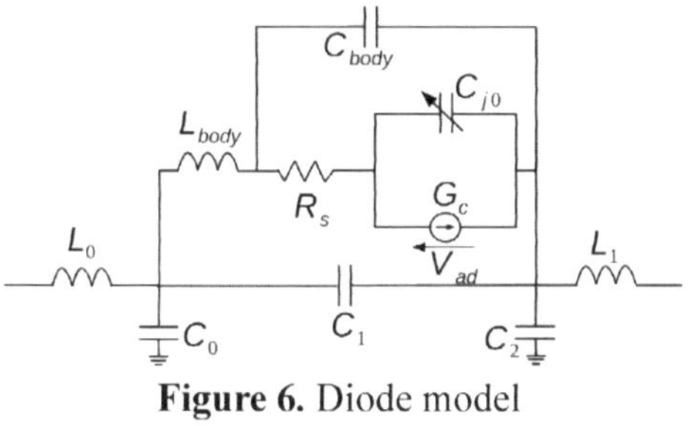

Figure 6. Diode model

This comparison is done at the particular diode polarization voltage of 84 mV. This voltage in simulation gives the best efficiency with a load of 5.7 kΩ and a lossless impedance matching between the generator and the circuit. A difference in phase has been noticed, which could be explained by the block-to-block co-simulation. This process, hence does not take into account the coupling effect in practice between the elements.

3.2. Simulation

The simulated results of the rectenna are compared to our previous results [5] (Figure 7). In both simulations, we use S-parameters of the simulated CPW elements of the DC filter and the input-matching network. We simulate the full rectenna at 2.45 GHz and choose a L-matching impedance network realized in CPW technology. This matching technique is simple, uses fewer components and avoids the use of component mounted in surface. In this work, we use the antenna impedance in the simulation. To do so, a S-parameter block is used which allows the rectifier to only see the antenna impedance while being powered by the generator (Figure 8).

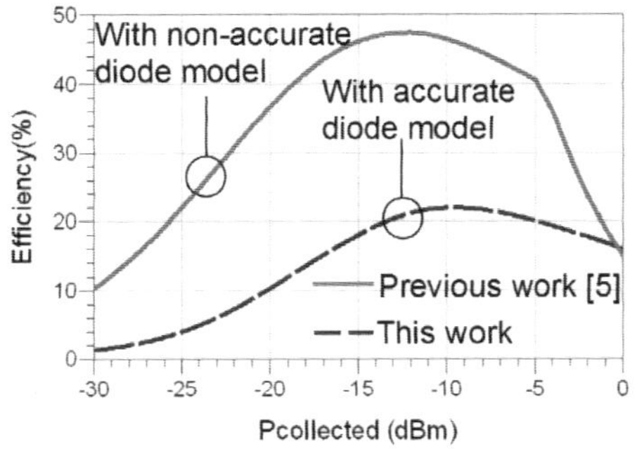

Figure 7. Simulated rectenna efficiency

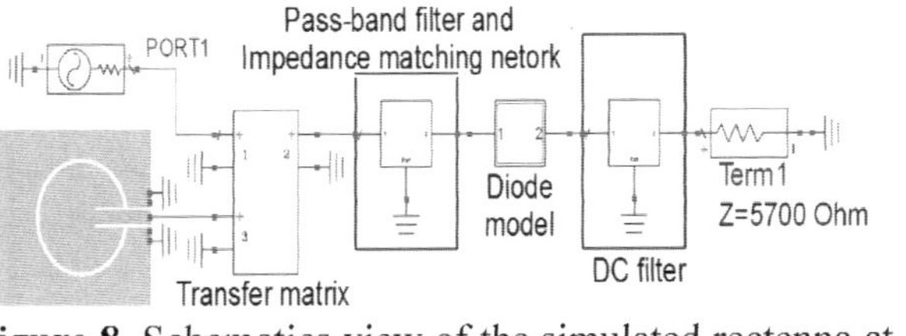

Figure 8. Schematics view of the simulated rectenna at 2.45 GHz

4. Conclusion

In this work a compact CPW-fed circular slot loop antenna is designed and measured. We show that such an antenna can be derived from a circular monopole antenna as a starting point. Those modifications reduce the bandwidth and the size of the antenna, but improve the gain at the desired frequency. This design is interesting because it has several advantages over the well-known patch antenna; such as, bad impedance matching at first higher order harmonics, larger bandwidth than a microstrip patch antenna, bi-directional radiation pattern in the opposite direction of space and antenna size reduction.

The simulation of a complete rectenna in CPW technology is performed with a more accurate diode model than our previous work [5]. The rectifier is composed of a single series diode HSMS7630. The diode is modeled through I/V characteristics and S-parameter fitting between diode model and measurement at low power level. The simulation of the rectenna shows rectifying efficiency of 10% at 2.45 GHz for P_{in}=-20 dBm and a resistive load R_{load}=5,7 kΩ.

Acknowledgment

This work received financial support from Reunion island Regional Council and Union European – European Regional Develoment Fund (FEDER)

References

[1] C. Perera, A. Zaslavsky, P. Christen and D. Georgakopoulos 2014 Context aware computing for the internet of things: A survey IEEE communications survey & tutorials vol. 16
[2] R. Simons 2001 *Coplanar Waveguide Circuits, Components, and Systems* ed John Wiley
[3] L. Xian 2012 Ultra-Wideband Antenna and design
[4] N. Zhu and R. W. Ziolkowski *2012* Metamaterial-inspired GPS L1 rectennas Antenna Technology (iWAT) IEEE International Workshop on p. 193–196.
[5] J.Rivière, A. Douyère, F. Alicalapa and J.-D Lan Sun Luk 2016 DC-pass filter design with notch filters supperposition for CPW rectenna at low power level IOP Conf. Ser. Mater. Sci. Eng. vol.120

IOP Conf. Series: Materials Science and Engineering **198** (2017) 012006 doi:10.1088/1757-899X/198/1/012006

Linear Vector Quantisation and Uniform Circular Arrays based decoupled two-dimensional angle of arrival estimation

Joseph D. Ndaw, Andre Faye, Amadou S. Maïga
Universite Gaston Berger, BP 234 Saint-Louis, Senegal
E-mail: andre.faye@ugb.edu.sn

Abstract. Artificial neural networks (ANN)-based models are efficient ways of source localisation. However very large training sets are needed to precisely estimate two-dimensional Direction of arrival (2D-DOA) with ANN models. In this paper we present a fast artificial neural network approach for 2D-DOA estimation with reduced training sets sizes. We exploit the symmetry properties of Uniform Circular Arrays (UCA) to build two different datasets for elevation and azimuth angles. Linear Vector Quantisation (LVQ) neural networks are then sequentially trained on each dataset to separately estimate elevation and azimuth angles. A multilevel training process is applied to further reduce the training sets sizes.

1. Introduction

Estimating the directions of arrival (DOAs) of narrowband signals impinging on an antenna array is an important issue in many applications such as radars and wireless communications. System performance and system capacity in noisy environment can be significantly improved with the knowledge of DOAs. An accurate characterisation of propagation environment and a precise localisation of mobile terminal are needed to optimise the link in modern wireless communications. For such purposes, DOA estimation is a mandatory technique especially in smart antenna technology[1]. The DOA is used to dynamically optimise the radiation pattern of the antenna array, making real time processing required in smart antenna technology [2].

The symmetry properties of a uniform circular array (UCA) allow 360° of coverage and uniform performance regardless of angle of arrival in the azimuth plane. Thus, a UCA array is an attractive configuration for DOAs estimation [3] and several algorithms were proposed to estimate DOAs using UCA.

Many publications on Artificial neural networks (ANN) are devoted to DOA estimations but most of them deal with 1D-DOA estimation. Radial Basis Function (RBF), Multi-Layer Perceptron (MLP) and Linear Vector Quantisation (LVQ) neural networks have been used for DOA estimation [4, 5, 6]. There are also publications reporting 2D-DOA estimations with neural networks (RBF, RBF-MLP and MLP) [7, 8, 9]. It is shown in these works that neural networks models outperform subspace based model in terms of speed of computation but computational complexity grows linearly with dimension [7].

In this work, a Linear Vector Quantisation neural network (LVQ-NN) based model is developed for real time 2D-DOA estimation through two separate 1D-DOA estimations.

IOP Conf. Series: Materials Science and Engineering **198** (2017) 012006 doi:10.1088/1757-899X/198/1/012006

2. Data model

Consider an M-element UCA with omnidirectional antennas uniformly distributed on the x-y plane. Source elevation angle, $\theta \in [0, \pi/2]$, and azimuth angle, $\phi \in [0, 2\pi]$, in a spherical coordinate system, are measured from the z and the x axis, respectively. If K narrowband incident plan waves $s_{i(i=1,2,...,M)}$ from directions $((\theta_1, \phi_1), (\theta_2, \phi_2), \ldots, (\theta_K, \phi_K))$ impinge on the array, the complex received signal x_m at the mth antenna is noise corrupted and is written as:

$$x_m(t) = \sum_{k=1}^{K} s_k(t) e^{-j\frac{2\pi a}{\lambda}\sin\theta_k \cos(\phi_k - \phi_m)} + n_m(t) \quad (1)$$

$m = 1, 2, \ldots, M$, λ denotes the wavelength. The couple of angles (θ_k, ϕ_k) is the 2D-DOA of the kth incident wave, $\phi_m = 2\pi\,(m/M)$ is the element angular position and n_m denotes the receiver noise of the mth element. The complex received signals are noise-corrupted and are written as:

$$[x(t)] = [A(\theta, \phi)] \cdot [s(t)] + [n(t)] \,. \quad (2)$$

$[x(t)]$, $[A(\theta, \phi)]$, $[s(t)]$ and $[n(t)]$ denote the vectors of antenna array outputs, array steering matrix, source signals and zero-mean spatially uncorrelated additive noise, respectively.

The array-processing algorithm uses the spatial covariance matrix ($\mathbf{R}$) i.e. cross-covariance information among the noise-corrupted signals. Let X be denoted as the noise corrupted data matrix composed of N snapshots of $x(n)$: $X(n) = [x_1(n)\; x_2(n) \ldots x_M(n)]$.

$$\mathbf{R} = \mathbf{A}E\left\{x(t)x^H(t)\right\} \approx \frac{1}{N}X^H X \quad (3)$$

$E\{\}$ denotes the statistical expectation and X^H the conjugate transpose of X. The input vector at the input layer of the LVQ network is the $\mathbf{R}$ matrix organized as a $2M^2$-dimensional vector.

3. LVQ network

An LVQ is a feedforward network with one hidden layer of neurones fully connected with the input layer. As a supervised method, LVQ uses known target output classification for each input pattern. The accuracy of classification depends on several factors. The number of input and output neurones are given by the dimensions of the input vectors and the number of predefined classes, the number of hidden neurones as well as the learning rate are chosen to get the best classification results. This can be done through an optimisation process [6].

4. Simulations and results

LVQ networks' ability to perform accurate 1D-DOA estimation within the 3-dB bandwidth of the radiation pattern was demonstrated in reference [6]. Computer simulations are based on a 9-element UCA. A 0.73λ radius is used to avoid coupling as reported in reference [10].

The spatial covariance matrix $\mathbf{R}$ is estimated and its elements are used to represent the signal at the array output. DOA estimation is made through a multilevel procedure over predefined sectors of the array field of view illustrated on figure 1. for elevation angles. In this case a 2-level procedure is applied. The first level, the classification step, gives a coarse location of the source within predefined subsectors of width $\delta\theta_1$ spanning the field of view $\Delta\theta$. The second level, the estimation step, gives a precise estimation of the DOA of the signal by classifying it within one of the predefined classes of a subsector of width $\delta\theta_2$.

Additionally, 2D-DOA estimation is made through two 1D simulations to reduce the training set size. Up to 98% training set reduction can be obtained for a 1° angle resolution [4]. This approach is of particular interest as 2D-DOA estimation with ANN requires very large training sets. A 1° angle resolution requires a 32 760-length training set for 2D-DAO estimation training while this length falls to 451 for two 1D-DOA estimations (91 for elevation and 360 for azimuth).

IOP Conf. Series: Materials Science and Engineering **198** (2017) 012006 doi:10.1088/1757-899X/198/1/012006

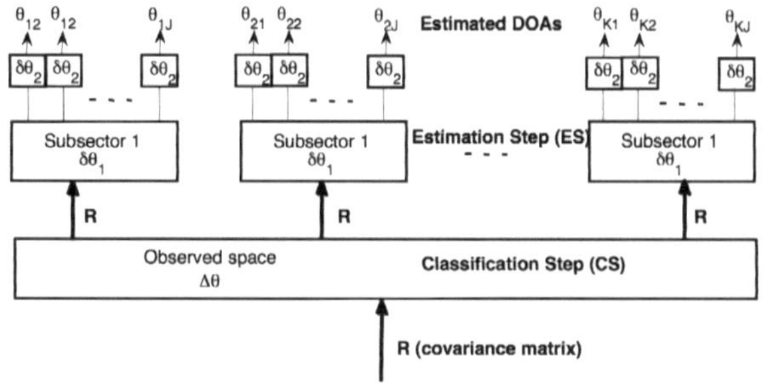

Figure 1. Sectorised field of view for multilevel LVQ model.

We first estimate the elevation angle (θ) from the array steering vector in which azimuth dependence is eliminated using UCAs circular symmetry. The array factor $A(\theta,\phi)$ of an M-element UCA is written as: $A(\theta,\phi) = \sum_i^M e^{j(2\pi/\lambda)a\sin\theta\cos(\theta_i-\phi_i)}$. The Anger-Jacobi expansion of the exponential term leads to: $A(\theta,\phi) = \sum_i^M \sum_{-\infty}^{+\infty} j^m J_m(kr\sin\theta)e^{jm(\phi-\phi_i)}$[1]. For an 9-element UCA array, the array factor can be approximated by the zero order Bessel function : $A(\theta,\phi) = MJ_0(kr\sin\theta)$. This form is used to build the $\mathbf{R}_\theta$ training set, used for elevation angle estimation. A second training set with no θ dependence, $\mathbf{R}_\phi$, for azimuth angle estimation is built with the estimated elevation angle $(\widehat{\theta})$: $x_{\phi i} = x_i^{(1/\sin\widehat{\theta})}$.

Angle estimation: Training sets are built with 1024 snapshots at a 20 dB signal-to-noise ratio (SNR) and a sampling step ranging from 2° in the first level to 0.25° in the last level. Confusion matrix rates are higher than 90% in the first level and simulation time is about one second with an 8 GB RAM computer. A 2-level model with $\Delta\theta = 32°$, $\delta\theta_1 = 6°$ and $\delta\theta_2 = 1°$ is used for elevation angle estimation while a 3-level model with $\Delta\phi = 360°$, $\delta\phi_1 = 60°$, $\delta\phi_2 = 10°$, $\delta\phi_3 = 2°$, was necessary for azimuth angle estimation. Angle estimation is based on a hundred runs and DOA is selected as the most probable estimated angle [6].

Table 1. Correlation Coefficient and RMSE for Elevation and Azimuth angles estimations.

Elevation			Azimuth		
	RMSE[θ]	r[θ]	$\hat{\theta}$	RMSE[ϕ]	r[ϕ]
Case 1	0.91°	0.9959	$\theta = 5°$	1.48°	0.9999
Case 2	0.61°	0.9979	$\theta = 10°$	2.43°	0.9997
Case 3	0.84°	0.9964	$\theta = 19°$	1.50°	0.9998

Estimation results: Several simulations were performed for elevation and azimuth estimation. Estimation performance of the LVQ-NN system can be expressed in terms of Pearson product-moment correlation coefficient, r, between reference values and network responses. The correlation coefficient indicates how well the estimated angles match the actual ones. If the correlation coefficient is close to one then the neural network has an an excellent predictive ability, whereas r close to zero indicates a poor performance of the network.

The LVQ model demonstrates good generalisation and 2D-DOA angle estimation capabilities as illustrated on table 1. Correlation coefficients obtained from linear fittings and root mean-square errors (RMSE) are given for 3 elevation and 3 azimuth angles estimations. Worst correlation coefficients (0.9959 and 0.9994) for elevation and azimuth angles, respectively, are concordant with (0.9987 and 0.9997) given in reference [7] for MLP neural networks.

Estimation errors: Estimation error at a fixed elevation/azimuth angle is noise dependent as shown on figure 2. A comparison between error rates at different elevation angles is shown on figure 3. Estimation error decreases as SNR increases. Error rates are high (60% at 20 dB SNR) at the bandwidth limit ($\theta = 27°$) but negligible toward the center of the beam ($\theta = 12°$). Error is about 1° and 2° for elevation and azimuth angles, respectively.

5. Conclusion

An accurate and fast LVQ neural network-based model for 2D-DOA estimation is presented. The developed model decouples elevation and azimuth angles estimation. It takes advantage of the symmetry properties of UCA array to build elevation or azimuth dependent training sets. Decoupling elevation and azimuth angle led to smaller training sets and a faster process for two 1D-DOA searches. Angle estimate errors are 1° and 2° for elevation and azimuth angles. The advantages of this model are the execution speed and a small angle estimation error that can be reduced by increasing the number of model levels.

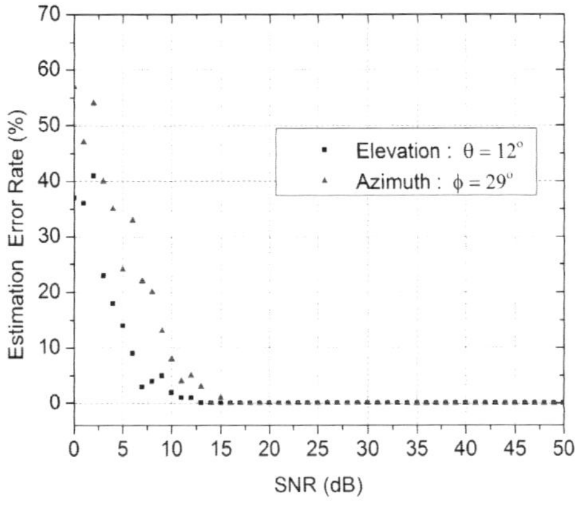

Figure 2. Estimated error rates for elevation and azimuth angles based on a 100 runs.

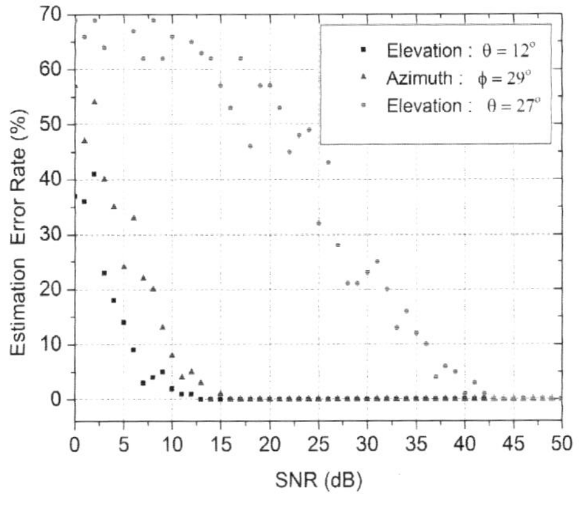

Figure 3. Estimated error rates and model limits shown at $\theta = 27°$ based on a 100 runs.

Acknowledgment

This work was supported by Universite Gaston Berger through projects : CEA-MITIC, VE.

References

[1] Balanis C. A. 2006 *Antenna theory: analysis and design* (Wiley-Interscience, New York)
[2] Liberty J. C. Jr and Rappaport T. 1999 *Smart Antennas For Wireless Communications*, (Prentice Hall PTR)
[3] Roald G and Rogier H 207 *EEE Trans. Antennas Propag.* **55**, p. 984
[4] El Zooghby, A H, Christodoulou C G, and Georgiopoulos M 2000 *IEEE Trans. Antennas Propag.***48** 768
[5] Gotsis K A, Siakavara K, and Sahalos J N 2009 *IEEE Trans. Antennas Propag* **57** 1399
[6] Faye A, Youm A B, and Ndaw J D 2013 5th *Int. Conf. on Computer Intelligence, Communication Systems and Networks (CICSyN)*, p 245
[7] Agatonovic M, Zoran S, Ivan M, Nebojsa D, Leen S, Thomas Z and Bratislav M 2013 *Progress In Electromagnetics Research* **137**, 741
[8] Fonseca N, Coudyser M, Laurin J J, and Brault J J 2010 *IEEE Trans. Antennas Propag.*, **58** 357
[9] Agatonovic M, Stankovic Z and Milovanovic B 2012 6th *European Conf. Antennas and Propag. (EUCAP)* p 1
[10] Das S and Mandal D 2011 *IEEE Indian Antenna Week* p 1

IOP Conf. Series: Materials Science and Engineering **198** (2017) 012007 doi:10.1088/1757-899X/198/1/012007

Joint Channel and Phase Noise Estimation in MIMO-OFDM Systems

I M Ngebani[1], J M Chuma[2], I Zibani[1], E Matlotse[1], and K Tsamaase[1]

[1] Faculty of Engineering Technology, Department of Electrical Engineering, University of Botswana, Private Bag 0061, Gaborone, Botswana

[2] Department of Electrical Engineering, Botswana International University of Science and Technology, Private Bag 24, Palapye, Botswana

E-mail: iboz55@gmail.com

Abstract. The combination of multiple-input multiple-output (MIMO) techniques with orthogonal frequency division multiplexing (OFDM), MIMO-OFDM, is a promising way of achieving high spectral efficiency in wireless communication systems. However, the performance of MIMO-ODFM systems is highly degraded by radio frequency (RF) impairments such as phase noise. Similar to the single-input single-output (SISO) case, phase noise in MIMO-OFDM systems results in a common phase error (CPE) and inter carrier interference (ICI). In this paper the problem of joint channel and phase noise estimation in a system with multiple transmit and receive antennas where each antenna is equipped with its own independent oscillator is tackled. The technique employed makes use of a novel placement of pilot carriers in the preamble and data portion of the MIMO-OFDM frame. Numerical results using a 16 and 64 quadrature amplitude modulation QAM schemes are provided to illustrate the effectiveness of the proposed scheme for MIMO-OFDM systems.

1. Introduction

Multiple-input multiple-output orthogonal frequency division multiplexing (MIMO-OFDM) systems are currently been deployed in the latest forth generation (4G) wireless systems. Similar to the OFDM case, phase noise in MIMO-OFDM systems leads to common phase error as well as inter-carrier interference due to loss of orthogonality among the subcarriers. CPE estimation schemes for MIMO-OFDM were derived in [1] and [2] by assuming that the channel state information (CSI) is perfectly known at the receiver and only for a case where the receiver antennas are affected by the phase noise. Together with CSI estimation the problem of phase noise estimation especially for multiple antennas is only rarely discussed [3]. To the best of the author's knowledge, no publication exists in the up to date literature about the joint channel and phase noise estimation and compensation in MIMO-OFDM systems where both the transmit and receive antennas are equipped with an independent oscillator.

2. MIMO-OFDM with Phase Noise

Fig. 1(a) and Fig. 1(b) show the block-diagrams of the MIMO-OFDM transmitter and receiver, respectively. In general, a MIMO-OFDM transceiver has N_t transmit antennas and N_r receive antennas, where each antenna is equipped with its own independent oscillator. The received

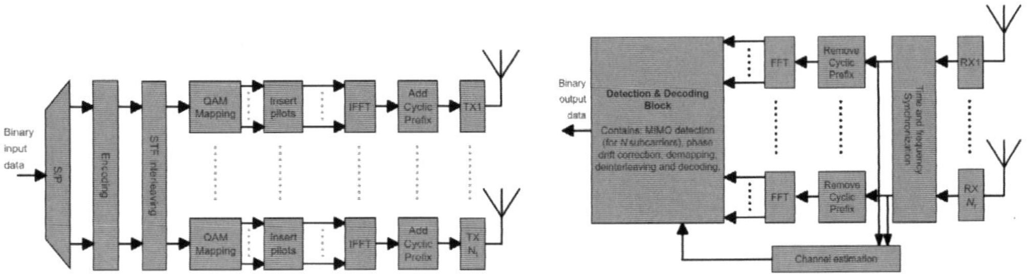

(a) MIMO-OFDM Transmitter (b) MIMO-OFDM Receiver

Figure 1: MIMO-OFDM Tranceiver

time domain signal at the q-th receive antenna in the presence of phase noise can be expressed as

$$y_q(n) = e^{j\theta_q^{[r]}(n)} \sum_{m=1}^{N_t} h_{q,m}(n) \star e^{j\theta_m^{[t]}(n)} x_m(n) + w_q(n) \tag{1}$$

where $x_m(n)$ is the signal transmitted by the m-th transmit antenna, $h_{q,m}$ describes the time domain channel impulse response between the mth transmit antenna and the qth receive antenna, $\theta_q^{[r]}(n)$ and $\theta_m^{[t]}(n)$ are the phase noise processes experienced at the q-th and m-th receive and transmit antennas, respectively. $w_q(n)$ is the additive white Gaussian noise and is assumed to be circularly symmetric complex Gaussian $\left(w_q(n) \sim \mathcal{N}(0, \sigma^2)\right)$. The operator $\star$ represents the convolution operation. Assuming that the cyclic prefix (CP) has been perfectly removed and after performing the discrete Fourier transform (DFT) demodulation at the receiver, the signal received at the q-th receive antenna at the k-th sub-carrier is given by

$$\begin{aligned} Y_q(k) &= \sum_{j=0}^{N_c-1} P_q^{[r]}(k-j) \sum_{m=1}^{N_t} H_{q,m}(j) \sum_{i=0}^{N_c-1} P_m^{[t]}(j-i) X_m(i) + W_q(k) \\ &= \sum_{m=1}^{N_t} H_{q,m}(k) \underbrace{P_q^{[r]}(0) P_m^{[t]}(0)}_{CPE_{q,m}} X_m(k) + \underbrace{\sum_{m=1}^{N_t} \left[\sum_{\substack{j=0 \\ j \neq k}}^{N_c-1} P_q^{[r]}(k-j) H_{q,m}(j) \sum_{\substack{i=0 \\ i \neq j}}^{N_c-1} P_m^{[t]}(j-i) X_m(i) \right]}_{ICI_{q,m}} \\ &\quad + W_q(k) \end{aligned} \tag{2}$$

In equation (2) above $P_q^{[r]}(k)$ and $P_q^{[t]}(k)$ are the receiver and transmitter phase noise Fourier coefficients given by $P_q^{[r]}(k) = \frac{1}{N_c} DFT\{e^{j\theta_q^{[r]}(n)}\}$, $P_m^{[t]}(k) = \frac{1}{N_c} DFT\{e^{j\theta_m^{[t]}(n)}\}$, respectively. $H_{q,m}$ is the frequency domain channel impulse response between the q-th receive and m-th transmit antennas.

2.1. Channel Estimation

Fig. 2 shows the arrangements of pilots and data symbols in a MIMO-OFDM frame that can be used for effective channel and phase noise estimation. In the preamble section of the MIMO-OFDM frame, a transmitter transmits known pilots in two consecutive time slots while the rest of the transmit antennas do not transmit. In the first instant, half of the sub-carriers transmit nulls and the other pilots. In the second time instant the roles of pilot and null

IOP Conf. Series: Materials Science and Engineering **198** (2017) 012007 doi:10.1088/1757-899X/198/1/012007

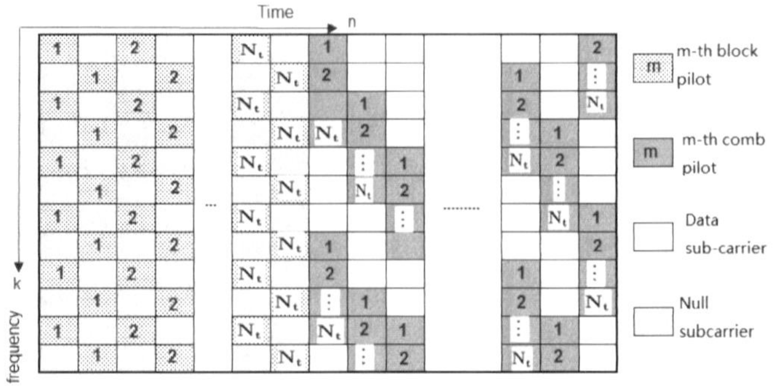

Figure 2: Pilot structure

sub-carriers are interchanged so that the channel estimates can be obtained at all sub-carriers via the use of the least squares (LS) method. The data section of the MIMO-OFDM frame employs a comb structure such that each sub-carrier carries data from a different antenna thus effectively implementing a frequency division multiplexing (FDM) scheme. To begin the initial channel estimation, let $S^p_{m,1}$ denote the set of pilot indices in transmit antenna m during the first time instant whilst $S^p_{m,2}$ denotes the set of indices during the second transmission instant. We note that $|S^p_{m,1} \cup S^p_{m,2}| = N_c$, where N_c is the number of sub-carriers. Similarly $S^z_{m,1}$ and $S^z_{m,2}$ denotes the set of null sub-carriers of the m-th transmit antenna in the first and second transmissions, respectively. Let the received signal at the q-th receive antenna when transmit antenna m transmits be given by $Y^b_{q,m}(k)$ whilst the transmitted pilot sub-carrier is given by $X^b_m(k)$. The estimated channel coefficient if we apply the LS method is then given by $\hat{H}^b_{q,m}(k) = \frac{Y^b_{q,m}(k)}{X^b_m(k)}, \quad k \in S^p_{m,1} \cup S^p_{m,2}$. Since we do not know the variance of the sum of the ICI and AWGN, we assume it to be uncorrelated and Gaussian.[1] We estimate the variance by calculating the average energy of the null sub-carriers as follows $\sigma^2_{q,m} = \frac{1}{N_z} \sum_{k \in S^z_{m,1} \cup S^z_{q,m}} |Y^b_{q,m}(k)|^2$, where N_z denotes the cardinality of the set $S^z_{m,1} \cup S^z_{m,2}$.

2.2. CPE Estimation

A relative $CPE_{q,m}$ for the data portion can be calculated from the channel estimates that would have been obtained from the preamble by applying the least squares to the comb structure of pilot arrangements in the data portion. This relative $CPE_{q,m}$ is given by

$$CPE_{q,m} = \frac{\sum_{k \in S^c_m} Y^c(k) \hat{H}^*_{q,m}(k) X^*_m(k)}{\sum_{k \in S^c_m} |\hat{H}_{q,m} X_m(k)|^2} \tag{3}$$

The channel estimate as seen by the data portion of the frame is relative to the preamble section and is given by $\hat{\hat{H}}_{q,m} = \frac{\hat{H}_{q,m}}{CPE_{q,m}}$. Applying MIMO to sub-carrier index k, the signal received at all the antennas can is given by $\mathbf{Y}(k) = \hat{\hat{\mathbf{H}}}(k)\mathbf{X}(k) + \mathbf{V}(k)$, where $\mathbf{Y}(k) = [Y_1(k) \quad Y_2(k) \cdots Y_{N_r}(k)]^T$ and $\mathbf{X}(k) = [X_1(k) \quad X_2(k) \cdots X_{N_t}(k)]^T$

An estimate of the transmitted symbols $X(k)$ can be obtained by using MMSE equalization as $\hat{\mathbf{X}}(k) = \mathbf{C}^H(K)\mathbf{Y}(K)$. The MMSE equalizing matrix $\mathbf{C}(k)$ is given by

$$\mathbf{C}(k) = \left(\hat{\hat{\mathbf{H}}}(k) \hat{\hat{\mathbf{H}}}^H(k) + \mathbf{\Sigma}(k) \right)^{-1} \hat{\hat{\mathbf{H}}}(k) \tag{4}$$

[1] In reality the ICI in MIMO-OFDM systems exhibits spatial correlation and is not Gaussian.

The matrix $\mathbf{\Sigma}$ denotes the combined AWGN and ICI noise covariance matrix.

3. Simulation Results

To evaluate the performance of our technique, it was assumed that between every transmit antenna m and every receive antenna q, there is a complex SISO channel impulse response $h_{q,m}(k)$ of length $L+1$ where $L=7$. That is $\mathbf{h}_{q,m} = [h_{q,m}(0), \ldots, h_{q,m}(L)]^T$. All channels have the same channel order L and the channel energy is normalized so that $\sum_{k=0}^{L} E|h_{q,m}(k)|^2 = 1 \quad \forall q, m$. The number of subcarriers used is equal to $N = 512$, and the modulation schemes used are 16-QAM and 64-QAM. The simulation makes use of 2 transmit and 2 receive antennas. The number of pilot tones per transmitter is equal to 4.

3.1. Results and Discussion

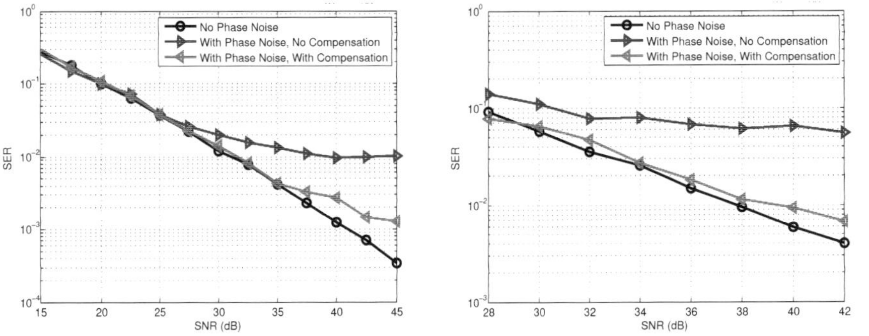

(a) SER for 16-QAM-MIMO-OFDM (b) SER for 64-QAM-MIMO-OFDM

Figure 3: SER Graphs for 16-QAM and 64-QAM Modulations for transmitter and receiver phase noise standard deviations of 1°

Fig. 3(a) and 3(b) show the symbol error rate (SER) results for 16-QAM and 64-QAM modulation schemes, respectively. The results show that the proposed scheme is very effective in estimating and compensating for phase noise. Results for 64-QAM shows that the higher the modulation scheme the more sensitive it is to the phase noise impairment. However, the results show that the proposed scheme is also effective for higher modulation orders.

4. Conclusion

In this paper the problem of joint estimation of channel and phase noise in a MIMO-OFDM system where both the receive and transmit antennas are equipped with independent oscillators was tackled through the use of a novel placement of pilots and nulls in the preamble and data portions of the MIMO-OFDM frame. Simulation results show that the proposed scheme is quite effective in jointly estimating the channel and the phase noise impairment. Furthermore, the effects of phase noise become much more severe as the modulation order is increased.

5. References

[1] T. C. W. Schenk, X.-J. Tao, P. F. M. Smulders, and E. Fledderus, "On the influence of phase noise induced ici in mimo ofdm systems," *IEEE Communications Letters*, vol. 9, no. 8, pp. 682–684, Aug 2005.
[2] K. Nikitopoulos and A. Polydoros, "Decision-directed compensation of phase noise and residual frequency offset in a space-time ofdm receiver," *IEEE Communications Letters*, vol. 8, no. 9, pp. 573–575, Sept 2004.
[3] H. Minn, N. Al-Dhahir, and Y. Li, "Optimal training signals for mimo ofdm channel estimation in the presence of frequency offset and phase noise," *IEEE Transactions on Communications*, vol. 54, no. 10, pp. 1754–1759, Oct 2006.

IOP Conf. Series: Materials Science and Engineering **198** (2017) 012008 doi:10.1088/1757-899X/198/1/012008

Impact of in-band interference on a wake-up radio system in wireless sensor networks

J M Lebreton, N M Murad and R Lorion

LE2P Laboratory, University of La Reunion, Saint-Pierre, France

E-mail: jean.lebreton@univ-reunion.fr

Abstract. The energy efficiency of Wireless Sensor Networks (WSNs) is considerably improved with Wake-up Radio (WuR) systems. However, their resilience to interference is often neglected in the literature. This might be an issue due to the proliferation of wireless devices and the growing field of internet of things. In this paper, we evaluate the impact of in-band interference from wireless devices on a WuR system. The approach proves that WuR systems are still performing well when coexisting with external wireless networks, even if the energy-efficiency is slightly reduced.

1. Introduction

Wireless Sensor Networks (WSNs) are exploited in numerous applications such as telemedicine, robotic, intrusion detection or environmental monitoring. WSNs are comprised of sensor nodes which can collect, process and transmit the data related to its application. Sensor nodes are operated through limited batteries and their replacement after depletion is usually not possible. An essential challenge is then to reduce the nodes energy consumption for extending the WSN lifetime.

Medium Access Control (MAC) protocols are designed to manage the wireless communication with the first goal being to avoid packets collision in the network. Various MAC protocols have been proposed in the literature to reduce energy wastage during data communication such as idle listening or overhearing [1]. Wake-up Radio (WuR) protocols have been recently proposed to replace traditional MAC protocols with the benefit of avoiding the aforementioned energy wastage. A WuR protocol essentially relies on the inclusion of a second radio into the node, the latter must be low-power or completely passive, and it is used to wake up the main radio when communication is required. The energy reduction through WuR protocols compared to conventional MAC protocols is well presented in the literature [2, 3].

However, WuR systems might be partially triggered by wireless devices communicating at the same frequency since their RF front end is typically an RF energy harvesting circuit. The harvested energy is only used by the WuR circuit here, but it is generally used for recharging the batteries and thus extending the WSN lifetime. The only work evaluating interference effects on WuR systems was performed in [3], but it was limited to the interference in a homogeneous network.

In this paper, we evaluate the interference from an external wireless network such as WiFi devices on the WuR systems. This approach aims to show the robustness of the WuR system while being interfered by existing wireless networks.

2. System model

WuR systems are defined by a specific hardware and protocol. The design of WuR receiver (WuRx) was presented in [4] and it should be added to the sensor nodes for enabling the WuR functionality. The WuRx is mainly composed of a compact rectenna array working at 2.45 GHz [5], an extremely low-power microcontroller (PIC24F16KA102) for validating the address included in the Wake-up Call (WuC) message, and a nano-power comparator which is used for increasing the WuRx sensitivity.

Our WuR protocol was first implemented in OMNeT++ with the MiXiM framework and its implementation details are given in [6]. This work has been further ported into the INET framework for more advanced simulations. The simulation framework gives reproducible and consistent results with the input parameters related to energy consumption and latency obtained by experimental measurements.

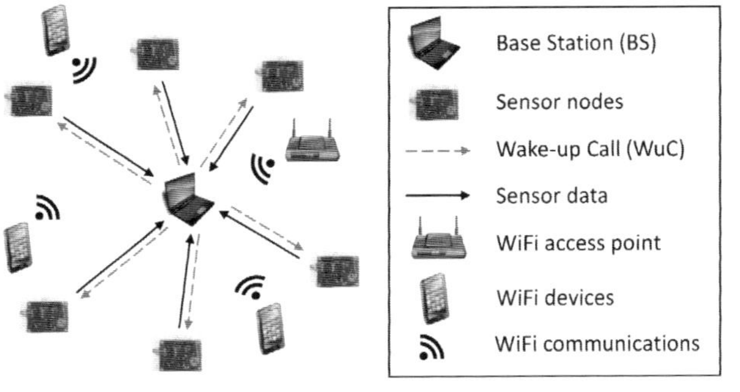

Figure 1. Network topology considered for the simulation.

The scenario considered in the simulations is illustrated in figure 1. The WuR network is composed of a Base Station (BS) and N sensor nodes equipped with a WuRx. Data are requested by the BS through the WuC message, which triggers the WuRx for waking up the addressed node. An interferer network is added through WiFi devices, all of them sending data to the WiFi access point. Both networks are operating at 2.45 GHz with a different modulation type, which is On-Off Keying (OOK) for the WuR system and Quadrature Amplitude Modulation (16-QAM) for the WiFi devices. The main simulation parameters are given in table 1.

Table 1. Simulation parameters.

Network	Parameter	Value
Common	Surface area	400 m^2
	Carrier frequency	2.45 GHz
	Simulation duration	30 min
WSN	Number of nodes	10
	Transmission power (BS)	100 mW
	Data size	100 bytes
WiFi 802.11g	Number of access point	1
	Data size	1 kbytes
	Inter packet arrival time	1 s
	Bit rate	2 Mbits/s
	Transmission power	100 mW

3. Results and Analysis

In order to evaluate the interference effects, four performance metrics are analysed here. First, the energy per bit is used to evaluate the energy-efficiency, and it is defined as the ratio of the total energy consumed by a node and the total number of correct bits received at the BS. Secondly, the average packet delivery ratio is used to evaluate the network reliability, which is calculated as the average number of packets received at the BS by the total number of requests (WuC messages) sent by the BS. Thirdly, the false alarm probability $P(FA)$ directly quantifies the interference impact on the WuRx. It is defined as the probability of detecting a wake-up signal when there is no wake-up signal in the channel. Fourthly, the Bit Error Rate (BER) gives an indication of the data link at the BS. It is calculated as the number of bit errors divided by the total number of bits received at the BS.

Several simulations are executed while incrementing the inter packet arrival time, which is the time between two consecutive packets reception at the BS. We varied this parameter from 200 ms to 20 s for considering a wide range of potential applications. Simulations are also performed for different numbers of interferers, varying from 0 to 6 in order to increase the interference traffic. Results are averaged for a consequent number of simulations and the 99% confidence interval is given.

Results regarding the energy per bit are depicted in figure 2. While the traffic rate in the WSN is relatively high, interferer nodes do not influence the WSN energy efficiency. However, when considering a low traffic rate in the WSN, the energy efficiency decreases with the number of interferers. This is mainly due to the small amount of energy consumed by the sensor nodes in this scenario, and the WuRx would be triggered numerous more times by interferers than it is fully activated by the BS. In the worst case, the energy per bit is multiplied by 2 when there are 6 interferers compared to the case without interferers.

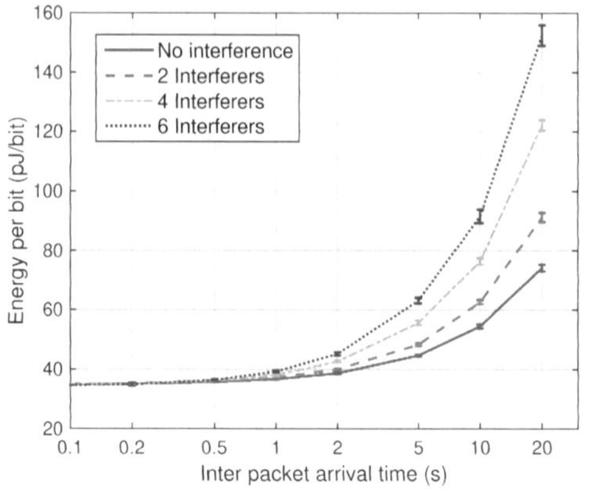

Figure 2. Energy per bit consumed by the nodes vs. the WSN inter packet arrival time and for different numbers of interferers.

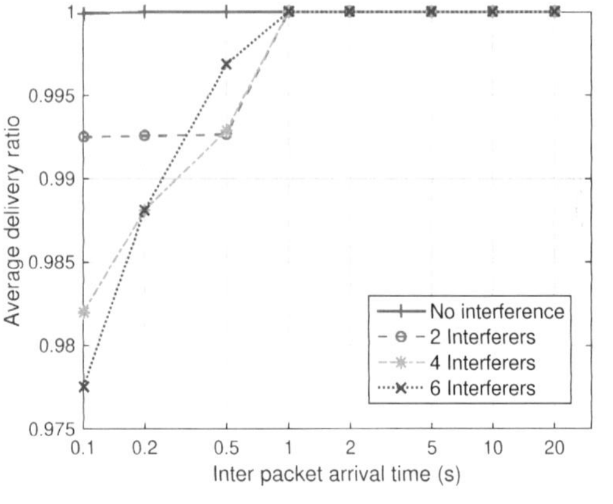

Figure 3. Average packet delivery ratio at the BS vs. the WSN inter packet arrival time and for different numbers of interferers.

The average packet delivery ratio is depicted in figure 3 for each scenario. First, every sensor node is successfully requested when there is no interference since all packets are received at the BS. Then, the packet delivery ratio globally decreases as the number of interferers increases, while considering WSN inter packet arrival time lower than 1 s. This packet loss is fairly acceptable because the average delivery ratio remains higher than 97.5 %.

Then, the results for the false alarm probability $P(FA)$ of the WuRx are represented in figure 4. $P(FA)$ is always lower than 5 % when there is no interference. However, $P(FA)$ varies from

8 % to 98 % while increasing the number of interferers and decreasing the WSN traffic. Those results clearly show the interference effects and they explain the decrease of energy efficiency. In fact, the rejection of WiFi interference is the reason behind the decrease of energy efficiency.

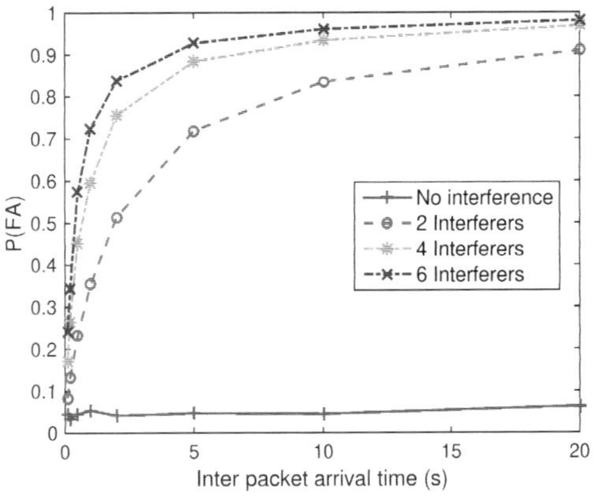

Figure 4. False alarm probability of the WuRx vs. the WSN inter packet arrival time and for different numbers of interferers.

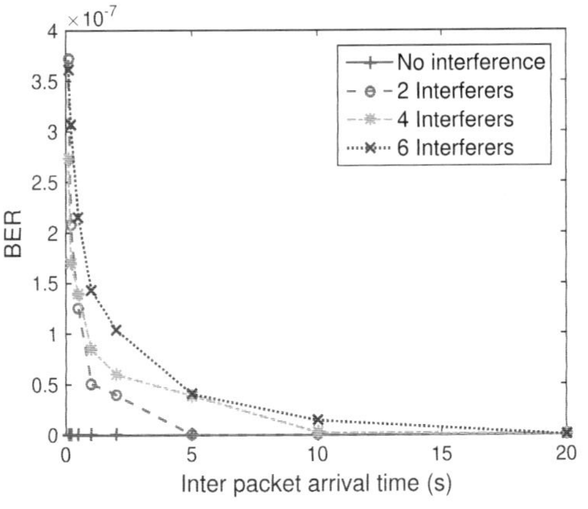

Figure 5. Bit Error Rate at the BS vs. the WSN inter packet arrival time and for different numbers of interferers.

Finally, the BER results are depicted in figure 5. As we can see, the BER increases with the number of interferers. Also, the BER is equal to 0 when there is no interference because the channel is modelled without noise. We did not model the noise for these simulations in order to only evaluate the effect of interference on the BER performance. In these scenarios, the BER varies between 0 and 4.10^{-7}, which is acceptable in most WSN applications.

4. Conclusion and Perspectives

This paper shows the effect of in-band interference on the performance of a WSN equipped with our WuR system. In the worst case where the data traffic is much heavier in the WiFi network than in the WSN, the energy efficiency can be halved. The main reason is the energy spent by the WuRx to reject the interference. However, the energy per bit in such circumstances remains much lower than the order of magnitude of typical WSNs without WuR systems. The network is still working correctly since the average packet delivery ratio remains higher than 97,5 %.

In future works, a more advanced WuR protocol or WuR system can be considered for avoiding or rejecting interference with less energy cost. The interference sources could also be exploited by the energy harvesting circuit for partially powering up the WuR system, or even totally if the harvested energy is sufficient. Finally, a practical implementation is going to be conducted for a comparison with the presented simulation results.

References

[1] Adhikari R 2104 *Journal of Network and Computer Applications* 41 pp. 488-504
[2] Spenza D et al. 2015 *IEEE Conference on Computer Communications (INFOCOM)* (Los Alamitos: IEEE Computer Society) pp. 522-530
[3] Oller J et al. 2015 *IEEE/ACM Transactions on Networking* 24 pp. 674-687
[4] Lebreton J et al. 2016 *IEEE Wireless Telecommunications Symposium (WTS)* (Los Alamitos: IEEE Computer Society) pp. 1-6
[5] Riviere J et al. 2014 *IOP Conf. Ser.: Materials Science and Engineering* 67 012009
[6] Lebreton J and Murad N 2015 *Proceedings of the 2nd OMNeT++ Community Summit* pp. 1-5

IOP Conf. Series: Materials Science and Engineering **198** (2017) 012009 doi:10.1088/1757-899X/198/1/012009

Arc Fault Detection & Localization by Electromagnetic-Acoustic Remote Sensing

C Vasile and C Ioana

UMR 5216 CNRS - Grenoble INP - Université Grenoble Alpes, 11 rue des Mathématiques, Grenoble Campus, BP46, F - 38402 SAINT MARTIN D'HERES Cedex

costin.vasile@gipsa-lab.grenoble-inp.fr, cornel.ioana@gipsa-lab.grenoble-inp.fr

Abstract. Electrical arc faults that occur in photovoltaic systems represent a danger due to their economic impact on production and distribution. In this paper we propose a complete system, with focus on the methodology, that enables the detection and localization of the arc fault, by the use of an electromagnetic-acoustic sensing system. By exploiting the multiple emissions of the arc fault, in conjunction with a real-time detection signal processing method, we ensure accurate detection and localization. In its final form, this present work will present in greater detail the complete system, the methods employed, results and performance, alongside further works that will be carried on.

1. Introduction

This current work intends to propose a case-study of an electrical arc fault detection and localization system, with emphasis on the detection and localization principles, and also on statistical performance evaluation. The physical principle behind this system is the simultaneous generation of the arc fault of electromagnetic waves, as well as sound (and low ultrasound) pressure waves. By exploiting this bidimensionality of the arc phenomena, the system (and the signal processing methods that it uses) manages to correctly assess the presence of the arc fault, as well as the distance at which it occurred.

In the first section, a brief description of previous and current embodiments of the system will be presented. Following, in section 2, we will describe the necessary signal processing in order to ensure robust detection of the arc fault. And in section 3, results and performances for one type of measurement will be presented, followed by our conclusions. The full paper version of the article will focus on the statistical evaluation of methods, in a wider range of configurations.

2. Experimental Configuration

The study of electrical arc fault has been widely treated in the scientific community [1]-[3], especially DC arc faults, that occur in photovoltaic power systems. This is the main focus of this work, as the type of arc faults that have been reproduced for this study are of the same nature.

IOP Conf. Series: Materials Science and Engineering **198** (2017) 012009 doi:10.1088/1757-899X/198/1/012009

Previous work on this issue has been carried out by the authors, with regards to the arc fault feature of generating electromagnetic and acoustic waves. This bidimensionality of the arc fault has been previously exploited in [4]-[6] with the use of a previous embodiment of a system composed from electromagnetic sensors in conjunction with an array of acoustic sensors. The electromagnetic sensor acts as an instantaneous detector and triggers the acoustic sensor data acquisition, which is then used for DTOA (Difference in Time of Arrival) estimation. The experimental system, with its remote sensing elements, is presented in figure 1.

An example of the setup is depicted in figure 2, with the arc fault source is located at a certain distance from the sensing system.

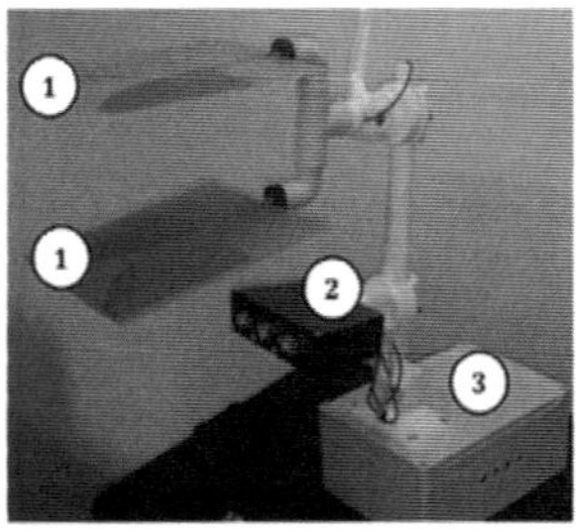

Figure 1. Current version of the detection and localization system.

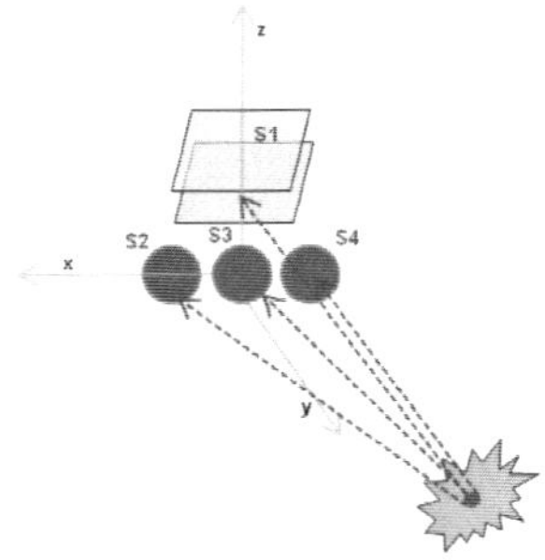

Figure 2. Arc fault source (yellow) & sensing system.

In the following section, we will present the signal processing methods used for DTOA estimation.

3. DTOA Estimation: Signal Processing

To better understand the nature of the localization problem, a first look at the three acoustic signals and the electromagnetic one is depicted in figure 3.

We see how the first sharp peak in the electromagnetic (blue) signal triggers the data acquisition, and how the delayed acoustic pressure wave reaches each of the sensors. The objective is to robustly detect those peaks and their moment of occurrence (Time of Arrival), in order to exploit the DTOAs for proper localization. The Short Time Fourier Transform is then used to analyze the frequency content of the acoustic signals, as we can see in figure 4, for one of the channels.

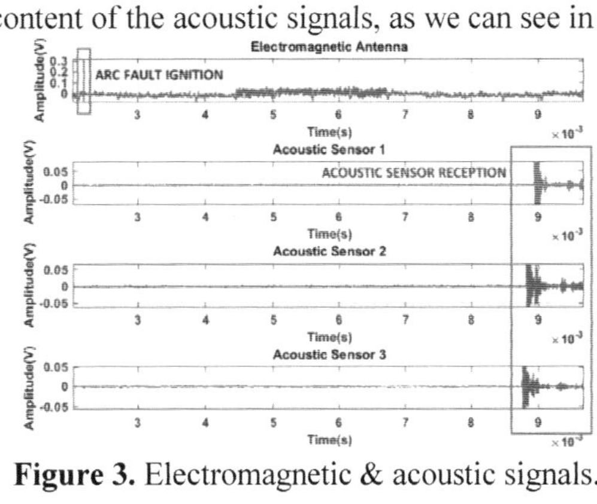

Figure 3. Electromagnetic & acoustic signals.

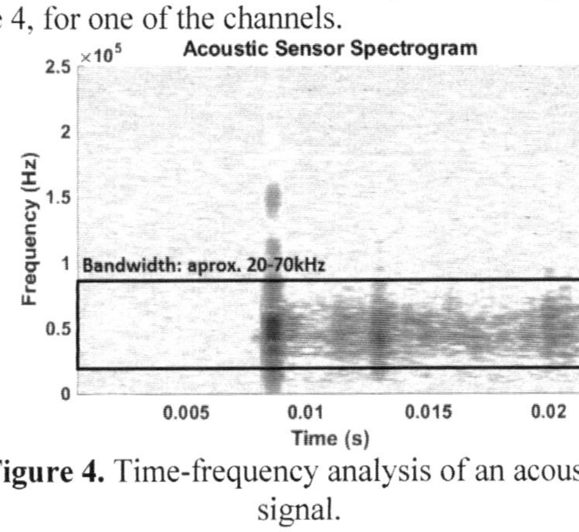

Figure 4. Time-frequency analysis of an acoustic signal.

We see, how the frequency bandwidth of interest is approximately between 20-70kHz.

The three proposed methods are:

- Peak of the envelope of the signal;
- RMS moving window;
- Statistical indicator moving window.

IOP Conf. Series: Materials Science and Engineering **198** (2017) 012009 doi:10.1088/1757-899X/198/1/012009

The choice for these methods is based on their simplicity, in order to prove the robustness of the physical detection principle: the electromagnetic-acoustic behavior of the arc fault facilitates the TOA estimation with different methods.

For a given signal, $s(t)$, its envelope, $R(t)$, is such that:

$$s(t) = R(t)\cos(\omega t + \phi(t)) \tag{1}$$

The RMS is such that, for a given sample k, of the discrete signal $s[n]$, the RMS moving window has the value at the time instant k:

$$RMS[k] = \sqrt{\frac{1}{L}\left(\sum_{i=k-L+1}^{k} s_i^2\right)} \tag{2}$$

, where L is the sample-width of the sliding window.

The statistical indicator $A[k]$ is the aggregate (arithmetic mean) of the L-sample-width window standard deviation and its sample kurtosis (similar to the RMS sliding window, of length L):

$$A[k] = \frac{\sigma_k + Kurt_k}{2} \tag{3}$$

The TOA's are given by:

$$TOA = \arg\max_n(Method[n]), Method \coloneqq R[n], RMS[n], A[n] \tag{4}$$

In order to properly localize the arc fault, we proceed to solve the geometric problem [4]-[6]:

$$\begin{cases} d_{SP_3} - d_{SP_2} = v \times DTOA_{32} \\ d_{SP_4} - d_{SP_3} = v \times DTOA_{43} \\ d_{SP_4} - d_{SP_2} = v \times DTOA_{42} \end{cases} \tag{5}$$

, where $d_{SP_i} = \sqrt{(x_{S_i} - x_P)^2 + (y_{S_i} - y_P)^2 + (z_{S_i} - z_P)^2}$, $i = \overline{1,4}$, $S_i(x_{S_i}, y_{S_i}, z_{S_i})$, $DTOA_{(i+1)i} = \Delta TOA = TOA_{i+1} - TOA_i$ is the difference of time of arrival (*DTOA*), TOA_i is the time of arrival at each acoustic sensor S_i, $i = \overline{1,4}$, and v is the speed of sound.

4. Results and Performances

We conducted tests in the following configurations (with the following arc fault source position):

I. X_{ARC} = 150cm, Y_{ARC} = 180 cm, Z_{ARC} = 0cm;
II. X_{ARC} = 180cm, Y_{ARC} = 180cm, Z_{ARC} = 0cm;
III. X_{ARC} = 510cm, Y_{ARC} = 0cm, Z_{ARC} = 0cm.

The following results were obtained and averaged over 5 measurements per configuration, for each of the three methods of TOA estimation (the errors are underlined):

Table 1. Envelope Peak Results

Conf.	X_{ARC}	Err.(%)	Y_{ARC}	Err.(%)
1	154.3	2.9	187.9	4.4
2	183.7	2.1	186.1	3.4
3	519.6	1.88	2.4	-

Table 2. RMS Sliding Window Results

Conf.	X_{ARC}	Err.(%)	Y_{ARC}	Err.(%)
1	153.2	2.1	184.4	2.44
2	183.5	1.95	182.9	1.61
3	512.7	0.53	3.2	-

Table 3. Statistical Indicator Results

Conf.	X_{ARC}	Err.(%)	Y_{ARC}	Err.(%)
1	161.3	7.4	196.5	9.1
2	187.2	4	196.8	9.3
3	523.6	2.5	4.4	-

From the results tables 1-3, it is noticeable how the Statistical Indicator is more sensitive to noise and more importantly, to window length (more valid estimates are obtained when the window length L $\rightarrow \infty$), therefore we obtained the less precise localization results for the Statistical Indicator sliding window. The best results were obtained for the Envelope Peak method, as it does not require the use of a moving window, and, in consequence, can indicate the exact instant (e.g. time sample, TOA) the arc fault occurs.

Seeing as the envelope peak method was the most precise, we conducted a Monte Carlo simulation analysis based on the measurements (using the 5-measurement-per-configuration mean and sample standard deviation), in order to ensure the degree of separation between the probability distribution function of the maximum value of the arc fault acoustic signal (its ignition) and the amplitude of the noise. The survival function visualization (figure 5) shows us how the probability of the noise amplitude rapidly goes to zero, while that of the arc fault remains at 1, thus the risk of false alarms is 0% and probability of detection is 100%. As expected, the distance of the arc fault source with regards to the sensing system influences the performances, in the sense that there is less and less margin between the range of amplitudes which ensure 0% probability of false alarm and 100% probability of detection, as the arc fault source is farther away from the sensing system.

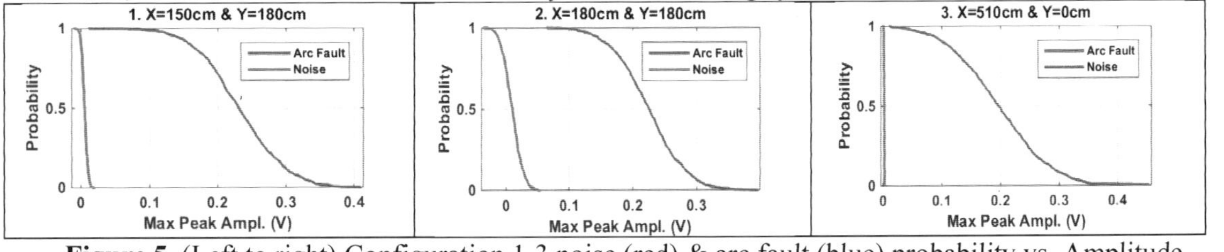

Figure 5. (Left to right) Configuration 1-3 noise (red) & arc fault (blue) probability vs. Amplitude threshold (survival function).

5. Conclusions

In this paper we have presented an updated version of an arc fault detection and localization system, which is based on the simultaneous emission of electromagnetic and acoustic waves from the arc fault. Experimental measurements have been carried out and different Time of Arrival estimation methods have been proposed, as well as the associated performances, which confirm the success of this approach. For the most promising of the methods, the envelope peak approach, we have constructed estimations of the statistical performances.

For future development, a more classical detection scheme will be investigated, such as the energy detector, with the corresponding enunciation of the detection problem in the shape of a binary hypothesis classification (arc fault vs. non-arc fault).

References

[1] J. Flicker, J. Johnson, "Electrical simulations of series and parallel PV arc-faults", IEEE PVSC, 3165 - 3172 (2013).

[2] J. Saulnier, J. Ghouili, "Detection of Electrical Arc Faults in a Distribution Network", IEEE CCECE, 571 - 574 (2007).

[3] J. Johnson et al., "Photovoltaic DC Arc Fault Detector testing at Sandia National Laboratories", IEEE PVSC, 003614-003619 (2011).

[4] A. Digulescu, M. Paun, C. Vasile, T. Petrut, D. Deacu, C. Ioana, R. Tamas, "Electrical arc surveillance and localization system based on advanced signal processing techniques", IEEE ENERGYCON, 426-430 (2014).

[5] C. Vasile, C. Ioana, A. Digulescu, I. Candel, "Electromagnetic and acoustic bimodality for the detection and localization of electrical arc faults", SPIE Advanced Topics in Optoelectronics, Microelectronics, and Nanotechnologies VIII, (2016).

[6] A. Digulescu, I. Candel, J. Dahmani, D. Deacu, C. Ioana, V. Gabriel, "Electric arc locator in photovoltaic power systems using advanced signal processing techniques", IEEE ELMAR, 129-132 (2013).

IOP Conf. Series: Materials Science and Engineering **198** (2017) 012010 doi:10.1088/1757-899X/198/1/012010

Comparison of RF Photonics-Based Beamformers for Super-Wide Bandwidth Phased Array Antennas

M. Belkin[1], A. Sigov[1], Y. Tyschuk[2], V. Golovin[2]

[1] Moscow State Technological University (MIREA), Moscow 119454, Russian Federation

[2] Sevastopol State University (SevSU) Sevastopol 299053, Russian Federation

E-mail: belkin@mirea.ru

Abstract. We demonstrate the NI AWRDE E-CAD tool-based simulation experiments to compare the three arrangements of photonic beam forming networks. The results confirm clearly the benefits of the proposed arrangement based on combination of multichannel fiber Bragg grating and switchable optical delay lines providing super-wide operating bandwidth and the better economical characteristics of microwave-band phased array antennas.

1. INTRODUCTION

As well known, antenna represents a basic unit of any on-air radio frequency (RF) system defining its communication range and fidelity. Nowadays, its exploitation in the communication and radar means as a phased array antenna (PAA) [1] results in remarkably increased footprint and operation flexibility thanks to electronic beam steering function that is fulfilled by a beam forming network (BFN). For modern communication and radar application [2] an extremely broad instantaneous bandwidth is required, so super broadband phase shifting or time shifting techniques must be utilized. On this way, RF and microwave photonics is a very attractive technology for PAA's BFN realization due to its super-wide operating bandwidth, immunity to electromagnetic interference, lightweight, and flexible system implementation [3]. So over the past years, a lot of RF photonic-based BFN architectures have been proposed including both phase shifting and time shifting techniques.

In process of R&Ds of such combined PAA including microwave and photonic units there was an issue which a computation technique is desired for their optimal modeling, analysis, and design. The essence is that for the accurate simulation of such complicated systems it is necessary to pay remarkable attention to their functioning in both spectral ranges. Now for the solution of this task, the approach referred to several computer-aided design (CAD) tools is in common use, because the existing optical and optoelectronic CAD tools (OE-CAD) are developed much weaker, than the CAD tools intended for modeling of microwave devices (E-CAD) [4]. In this regard, the conclusion has been drawn that the optimal way for increasing the accuracy of photonic circuits taking into account influence of their parasitic parameters in microwave band requires use of the high-power microwave E-CAD tool [5] working at symbolical level. Following the approach, we have developed by NI AWRDE computer tool a number of models for active optoelectronic elements, such as semiconductor laser [6, 7], p-i-n photodetector (PD) [8] and Mach-Zehnder modulator (MZM) [9].

Proceeding the works, in this paper the results of NI AWRDE-based simulation experiments to compare

IOP Conf. Series: Materials Science and Engineering **198** (2017) 012010 doi:10.1088/1757-899X/198/1/012010

the three versions of photonic BFN arrangements using photonic phase shifters, switchable optical delay lines, and our new arrangement based on a combination of multichannel fiber Bragg grating (FBG) and switchable optical delay lines are demonstrated.

2. BEAM FORMING NETWORK ARRANGEMENTS

Fig. 1,a shows the first photonic BFN arrangement for comparison that is a part of 16-element PAA transmitting channel. In this case, RF signals are converted by a laser to optical range and shared into 16 branches by optical coupler. Each branch consists of a tunable photonic phase shifter (P-PS), photodetector (PD), and antenna element. In the process of simulation experiment we used AWRDE equivalent circuit model including the model of directly modulated semiconductor laser [6], library coupler model of 1:16, the previously developed P-PS model [9], the model of PD [8], and ideal isotropic antenna model.

Fig. 1,b shows the second photonic BFN arrangement for comparison that is a part of 16-element PAA transmitting channel. In this case, 16 unmodulated lasers of different wavelengths λ1- λ16 are used. The part of the same RF signal is converted by corresponding Mach-Zehnder modulator (MZM) to optical range and shared into 16 branches by optical coupler. Each branch consists of a switchable optical delay lines (ODL). Then the delayed optical signals are converted into RF band by PD and emitted by antenna element. In the process of simulation experiment we used AWRDE equivalent circuit model including the model of continuous wave (CW) semiconductor laser [7], previously developed MZM model [9], library models of 1:16 couplers, switch, delay line, as such as the model of PD [8].

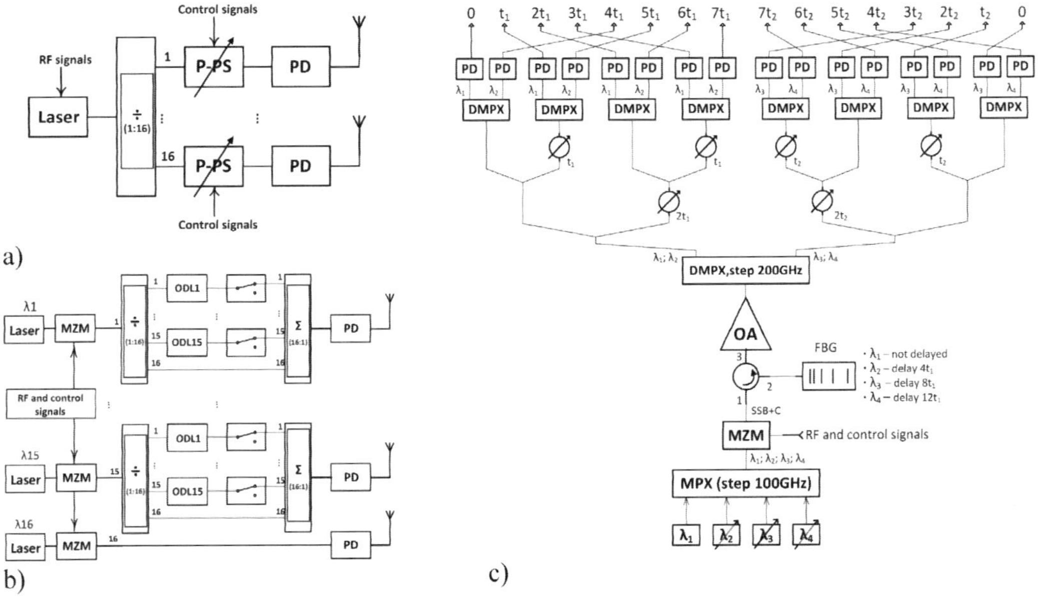

Fig. 1. 16-element RF photonic BFN based on photonic phase shifters (a), 16-element RF photonic BFN based on switchable optical delay lines (b) and 16-element RF photonic BFN based on a combination of multichannel fiber Bragg grating and switchable optical delay lines (c)

Fig. 1,c shows the third photonic BFN arrangement for comparison that is a part of 16-element PAA transmitting channel too. In this case, only 4 lasers of different wavelengths λ1- λ4 (3 tunable and 1 untuned) are used. Laser emissions are summarized in a spectral multiplexer (MPX), modulated in a common MZM by RF signal, and through optical circulator are input to 4-channel reflected FBG. The levels of corresponding delayed signals are recovered by an optical amplifier (OA), pairwise branched out, delayed once more by corresponding switchable ODL, branched out again by spectral demultiplexers (DMPX), converted into RF band by PDs and emitted by antenna elements.

IOP Conf. Series: Materials Science and Engineering **198** (2017) 012010 doi:10.1088/1757-899X/198/1/012010

3. NI AWRDE TOOL-BASED SIMULATION EXPERIMENTS

To evaluate phase shifter-based photonic beam former (Fig. 1,a) performances an equivalent circuit including the models of previously proposed microwave photonics components and built-in AWRDE's library models was developed. The simulation was performed by calculating the PAA's normalized radiation patterns (NRP) in the frequency band of 6-18 GHz using the well-known formulas [1]. Fig.2 shows examples of the calculation of the NRPs for the case of discrete phase delay selected in the center of a frequency band. As can be seen from the Figure, when a discrete phase shift $\Delta\varphi = 13°$ at the extreme frequencies of the band, there are significant errors $\delta\theta = 22°$ at the frequency of 6 GHz and $\delta\theta = 7°$ at a frequency of 18 GHz. From the Figure, one can make an unambiguous conclusion: to achieve acceptable error value the instantaneous band of RF signal must be narrower. As described above, the main advantage of the arrangement of Fig. 1,b is in the relative simplicity and straightforward implementation, as the scheme is a complete "optical" analogue of a standard microwave PAA scheme. To evaluate its characteristics an equivalent circuit including the models of previously studied microwave photonics components and built-in AWRDE's library models was developed. Modelling was also conducted by calculating the PAA's NRP in the frequency band 6-18 GHz. Fig. 3 depicts an example of calculation of the NRP in the upper frequency of the band for the photonic BFN based on switchable optical delay line with 3-ps increments. As one can see from the Figure, the photonic BFN is provided by a uniform spatial shift of the main lobe that is more than 40°. The same was observed on other frequencies of the band.

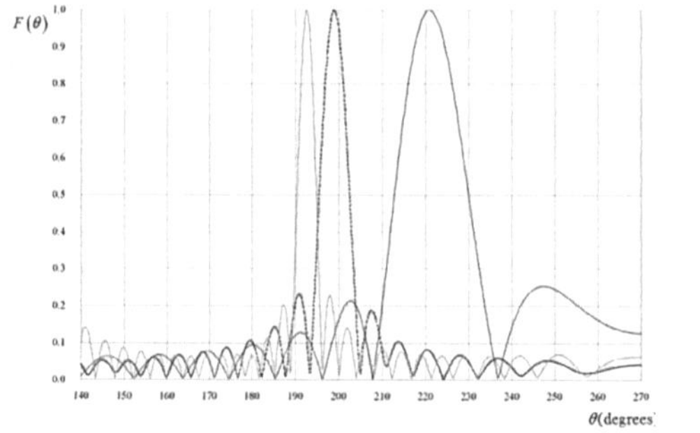

Fig. 2 Microwave photonics PAA NRPs at central frequencies of 6 GHz (right), 12 GHz (middle), and 18 GHz (left)

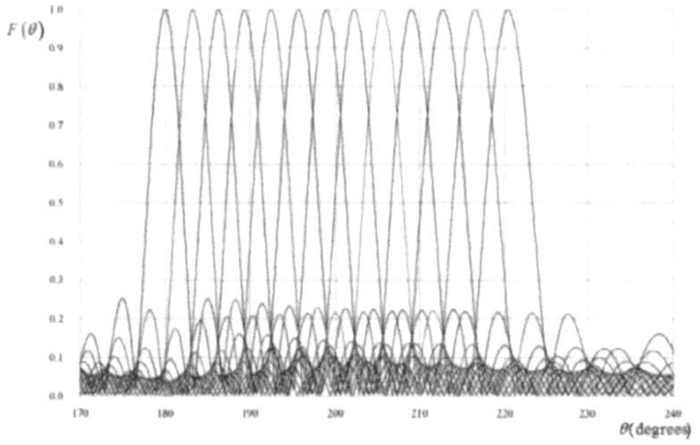

Fig. 3. Microwave photonics PAA NRPs at the frequency 18 GHz for photonic beam former using switchable delay lines

For simulating photonic beam former based on combining switchable delay lines and multichannel Bragg grating in two exiting modes of antenna elements the following reference data were taken: operating frequency band of PAA is 6-18 GHz; the antenna array has 16 non-directional antennas spaced at $0{,}5\lambda_{min}$ (corresponding to the maximum operating frequency of 18 GHz). Simulations are carried out for 2 cases: 1) at fixed microwave frequency 10 GHz when the delay time $\Delta t = 1$, 5, 10 and 15 ps; 2) at fixed time delay $\Delta t = 15$ ps when operating frequencies F= 6, 12 and 18 GHz.

The simulation results depict in Fig. 4 and Fig. 5, correspondingly. For comparison the same plots show the calculation results of ideal NRPs (point curves) [1]. The close coincidence proves the correctness of the proposed model. A clear output from the Figures is that the position of main lobe is unchanged, i.e. RF photonics-based beam former arrangement of Fig. 1,c is usable for super-wide bandwidth phased array antennas.

CONCLUSION

The comparison of the simulation results showed that the arrangement of Fig. 1,a, despite the greatest simplicity had a relative narrow instantaneous bandwidth (up to 300MHz) and so was unpractical for super wideband PAAs. The both residuary arrangements due to use so called true-time delay (TTD) technique [10] equally suit for precision high-speed steering of antenna beam in the band of 1-18 GHz with instantaneous bandwidth up to 5 GHz but the main benefit of Fig. 1,c's arrangement is in simplicity

IOP Conf. Series: Materials Science and Engineering **198** (2017) 012010 doi:10.1088/1757-899X/198/1/012010

and scalability. The last features one can clearly see from the Figs: for the realization of 16-channel BFN in the arrangement of Fig. 1,b as much as 16 lasers, 16 optical modulators, and 240 optical delay lines must be used. On the other hand, for the arrangement of Fig. 1,c only 4 lasers, one modulator, and 6 delay lines are needed.

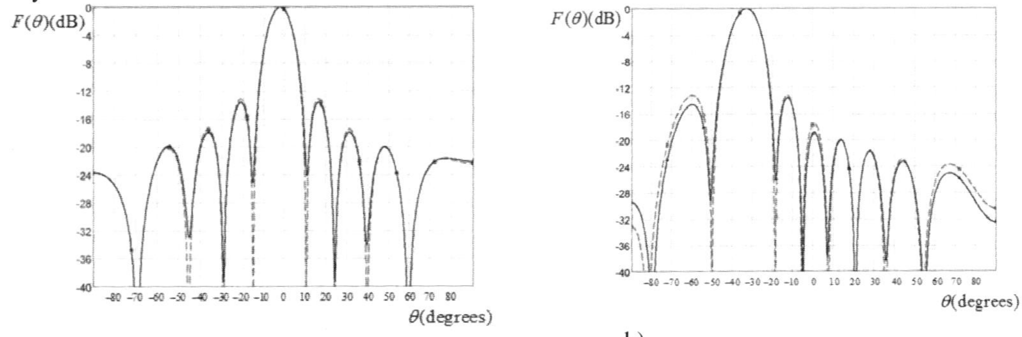

a) b)

Fig. 4. Radiation patterns of developed model of PAA at 10 GHz: a - Δt =1 ps; b - Δt = 15 ps

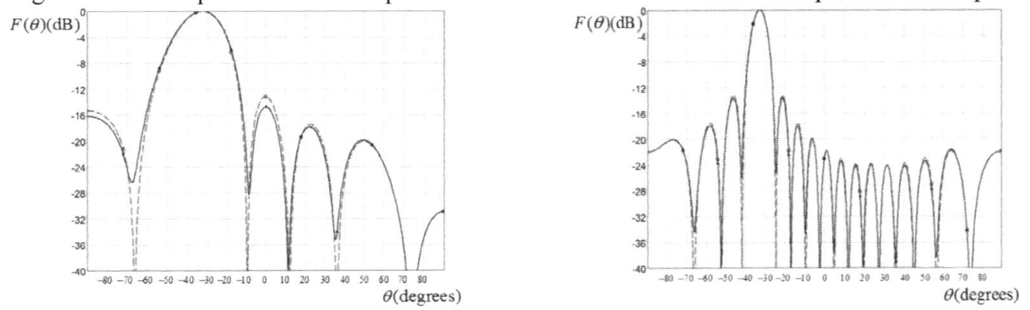

a) b)

Fig. 5. Radiation patterns of developed model of PAA at Δt = 15 ps: a - F = 6 GHz; b - F = 18 GHz

REFERENCES

[1] Mailloux R 2005 Phased Array Antenna Handbook *Artech House* p 496

[2] Bradsell P 1990 Phased arrays in radar *Electron. Commun. Eng. J.* pp 45-51

[3] Matthews P 1999 Practical photonic beamforming *Proc. IEEE Int. Topical Meeting on Microwave Photonics* pp 271-274

[4] Belkin M 2012 Multiscale Computer Aided Design of Microwave-Band P-I-N Photodetectors, pp. 231-250. *In book "Photodetectors" Ed. By S. Gateva – InTech, Chapter 6*

[5] Leijtens X, Le Lourec P, Smit M 1996 S-Matrix Oriented CAD-Tool for Simulating Complex Integrated Optical Circuits *IEEE J. of Selected Topics in Quantum Electronics* pp 257-262

[6] Belkin M, Belkin L, Sigov A, Iakovlev V, Suruceanu G, Kapon E 2011 Performances of Microwave-Band Analog Signal Transmission using Wafer-Fused Long Wavelength VCSELs *IEEE Phot. Technol. Lett.* pp 1463-1465

[7] Belkin M, Iakovlev V 2015 Microwave-band Circuit-level Semiconductor Laser Modeling *Proc. 9th European Modelling Symposium on Mathematical Modelling and Computer Simulation EMS2015* (Madrid Spain) pp 1-3

[8] Belkin M, Sigov A 2015 Circuit-Level Large-Signal Modeling of Microwave Bandwidth Photodetector *Proc. International Conference on Electromagnetics in Advanced Applications, ICEAA 2015* (Italy) pp 1587–1589

[9] Belkin M, Golovin V, Tyschuk Y 2016 The equivalent electrical model of the Mach-Zehnder electro-optical modulator at the CAD AWR Microwave Office *Proc. International Conference "Microwave & Telecommunication Technology" (CriMiCo'2016)* pp 2353 — 2359 (in Russian)

[10] Shaw A and Wilkins N 2010 Frequency Invariant Electro-Magnetic Source Location Using True Time Delay Beam Space Processing Proc *IEEE Intl. Symposium on Phased Array Systems and Technology*

IOP Conf. Series: Materials Science and Engineering **198** (2017) 012011 doi:10.1088/1757-899X/198/1/012011

Characteristics of Vivaldi antenna located on a cylindrical surface

Y V Yukhanov[1], A V Gevorkyan[1], T Y Privalova[2]

[1]Southern Federal University, Taganrog, Russia
[2]Southern Federal University, Rostov on Don, Russia
E-mail: yu_yukhanov@mail.ru

Abstract. The paper analyzes the directional properties of antipodal of Vivaldi antenna, located on the faces of the wedge, coupled with the two cylinders at the top and bottom. Two antenna patterns are considered. The first pattern has one antenna, located symmetrically on the edge of each face of the wedge, the second pattern has two of them. The comparative analysis of these antennas is done. The influence of the angle at the apex of the wedge and its size on the parameters of antennas is studied. The frequency and directional characteristics of antennas depending on their structural dimensions and their location are examined.

1. Introduction

«Vivaldi» antennas, including the ones in the composition of the active video pulse scanning antenna arrays (AR), are most widely used in a variety of ultra-wideband radio systems, communications, radar, and navigation. The development of possible options of constructing emitters of ultra short electromagnetic field pulses with one dimensional or two-dimensional scanning and given polarization characteristics is an urgent task [1-4].

In some cases, the antennas must be able to emit the shortest possible pulses. Such antennas include the lamellar structure in the form Vivaldi emitter with a power unit in the form of two cylinders spaced by a certain distance from each other, passing from one side into two metal emitter plates with the gap extending further [5]. The coaxial cable is connected to the area of the gap on the diametrically opposite side of the plates. The antipodal Vivaldi antenna has the same properties [6].
The purpose of this work is to study models of antipodal Vivaldi emitters, located on a cylindrical surface formed by the wedge conjugated with the cylinders at the top and bottom.

2. Statement of the problem

Antipodal Vivaldi antennas, made on a high-frequency dielectric with double-sided metallization, are located on a cylindrical surface formed by cylinder-wedge-cylinder conjugation (figure 1). This design can be used as a model of an aircraft wing or its part. Two variants of the antennas are considered. Two single antennas are located on the faces of the wedge symmetrically relatively to the *yz* plane (figure 1a) – antenna A_1. The planar Vivaldi antenna (figure 2) is orthogonal to the metal screen size 20mm*48mm - antenna A_2.
The wedge is made of a perfectly conducting material with through-cuts for antennas. Thus, the space inside the structure under the material from which the antennas are made (at the interval *L* (figure 1,b)) is filled with the vacuum.

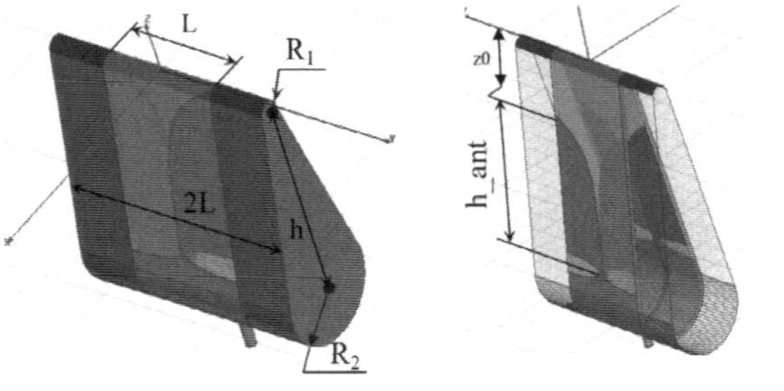

$a) z0 = 0$ b) $z0 \neq 0$

Figure 1. Vivaldi antennas A_1 on a cylindrical surface

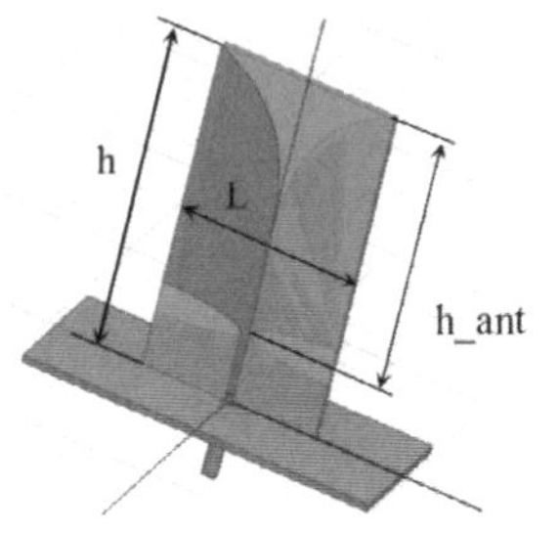

Figure 2. Vivaldi antennas A_2

It is necessary to study the directional properties of these antennas A_1 and A_2 depending on the geometry: length of the wedges' edges, cylinder radii, length h and width *L*.

3. Analysis of radiation characteristics of antenna

Numerical studies of antennas were carried out in HFSS software package [7]. Antenna designs were optimized for minimum VSWR and the maximum gain in the broadest possible operating frequency band. All antennas discussed below are made on a substrate Rogers RO4003(tm) with double-sided metallization 0.07 mm thick. The thickness of the substrate is 1mm. The width of the wedge in all antenna designs was taken twice as much as the width of the antennas, that is equal to 2L. For comparison, the planar Vivaldi antenna A_2 (figure 2) is made on the same substrate, and it is located perpendicularly to the metallic screen with dimensions 20mm*48mm. The dimensions of antenna A_2 are: width L, height from the main axis of the resonator h_ant, altitude to the screen h.

Let us study the influence of the surface's shape on the antenna's A_1 parameters with dimensions: L_1=24mm, h_ant=35mm, h=40mm from cylinders radii R_1 and R_2. The design parameter z0 (figure 1,b), determining the distance from the antenna aperture A_1 to plane z=0, is chosen equal to zero.
Figure 3 shows the dependence of VSWR from frequency for $R_1 = R_2 = 2$mm (curve ——), 4mm (curve — — —), 6mm (curve - - - -) and 8mm (curve ....). VSWR of A_1 is also shown here (curve — . . .).

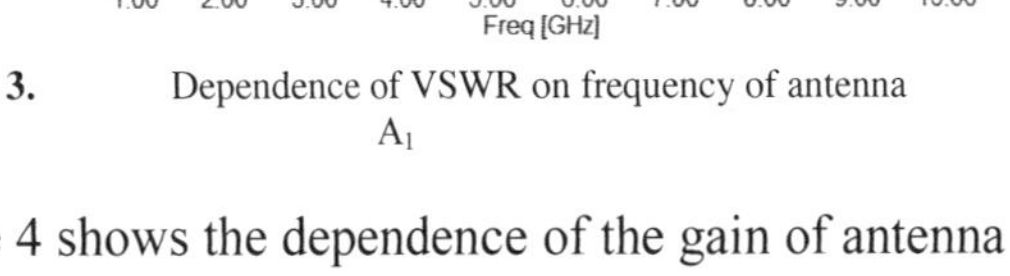

Figure 3. Dependence of VSWR on frequency of antenna A_1

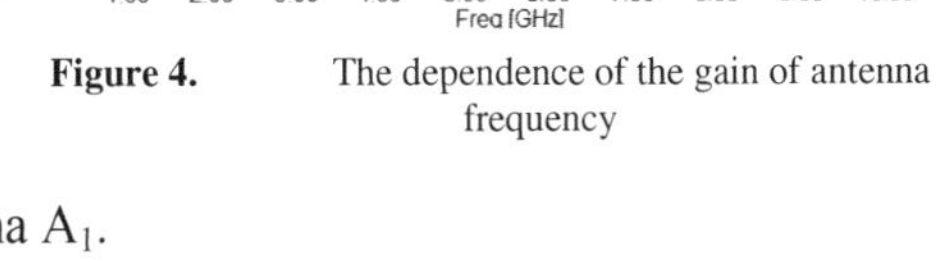

Figure 4. The dependence of the gain of antenna A_1 on frequency

Figure 4 shows the dependence of the gain of antenna A_1.

Type of antenna structure for $R_1 = R_2 = 8$mm is shown in the inset of figure 4. i.e. with increase of R_1 antenna design increasingly shields the aperture, which negatively affects the parameters of antenna (increases VSWR, decreases gain).
As it is seen from the analysis of figures the antenna with $R_1 = R_2 = 4$mm (see curves — — —) has the best parameters, in terms of the minimum VSWR in the broadest possible bandwidth and maximum gain.

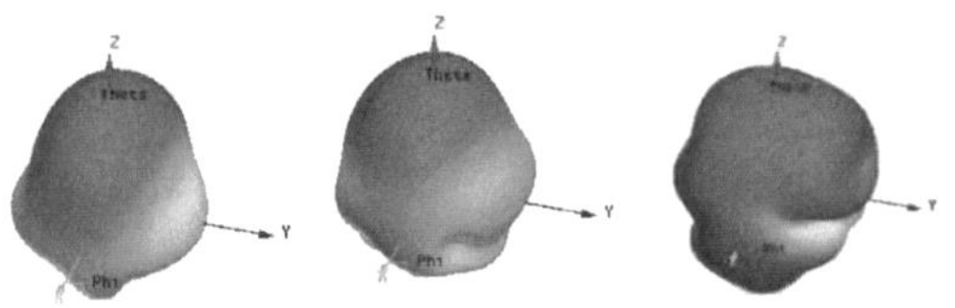

3.5GHz 7GHz 10GHz

Figure 5. 3D radiation pattern of antenna A_1

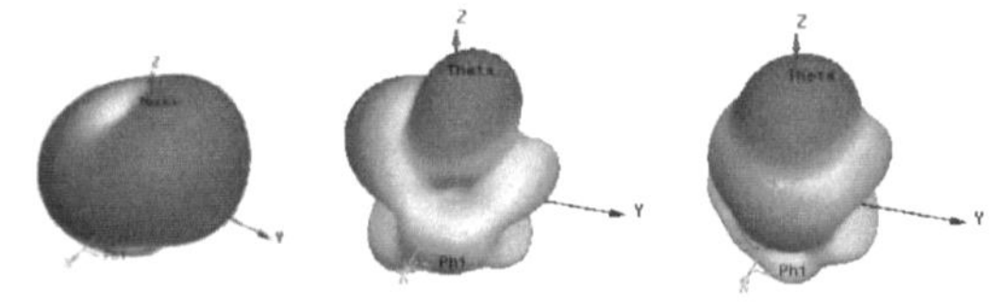

3.5GHz 7GHz 10GHz

Figure 6. 3D radiation pattern of antenna A_2

Figures 5 and 6 show 3D nonnormalized radiation patterns of antennas A_1 (figure 5) (with $R_1 = R_2 =$ 4mm) and A_2 (figure 6) on frequencies 3.5 GHz, 7 GHz and 10 GHz.

As can be seen, antenna A_1 in all considered parameters surpasses planar version of the same antenna (A_2).

Figures 7, 8 show the results of the analysis of influence of the radius $R_1 = 5$mm (curve ——), 10mm (curve – – –), 15mm (curve - - - -) and 20mm (curve ....), for R_2=2mm (diverging upward faces of the wedge (like that of a horn, see in the inset of figure 7)) on directional properties of A_1 antenna. Thus, to avoid obscuring the aperture, the distance z0 is chosen equal to radius R_1.

Figure 7 shows dependence of VSWR on frequency. As it is seen, the change in surface shape, in comparing with the previous case (figure 3), did not increase antenna's frequency band.

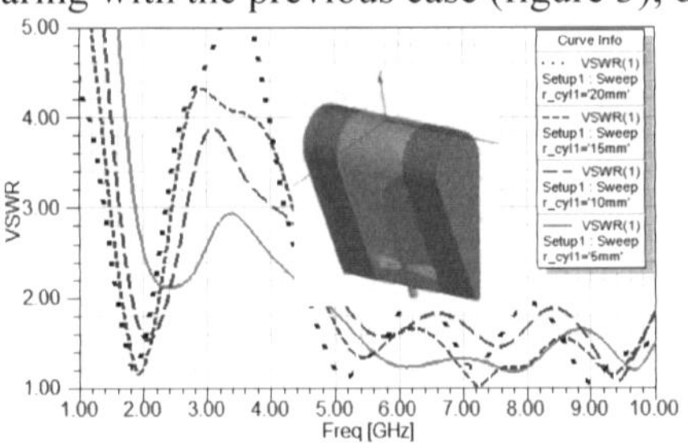

Figure 7. Dependence of VSWR on frequency of A_1 antenna for various $R_1 = 5mm; 10mm; 15mm$ and $20mm$

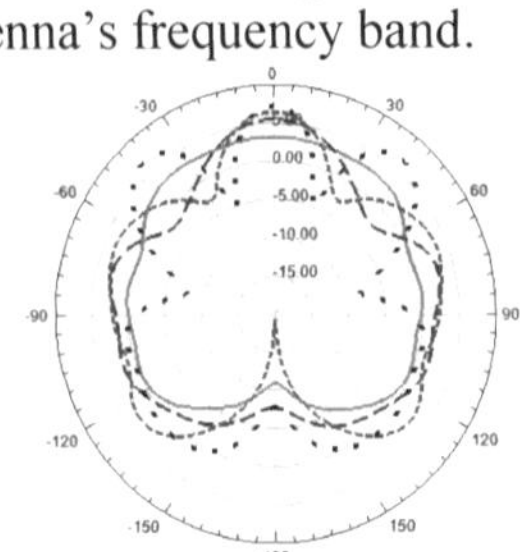

Figure 8. Radiation pattern of the antenna A_1 on frequency 10GHz for different R: $R_1 = 5mm; 10mm; 15mm$ and $20mm$

Moreover, the divergence of the wedge's edges (inset in figure 7 shows that the antenna design with the parameters $R_1 = 10$mm; R_2=2mm) leads to deterioration of antenna matching at frequencies 2.5GHz - 4.5GHz (VSWR increases with the increase of R_1).

Figure 8 shows the radiation pattern of antenna on H-vector plane at frequency of 10GHz for $R_1 =$ 5mm (curve ——), 10MM (curve – – –), 15MM (curve - - - -) and 20MM (curve ....) and for R_2=2mm.

Figures show that increase of radius R_1 does not lead to a simple narrowing of radiation pattern, as it is in a conventional horn antenna. In this case, the divergence of the wedge's edges at a distance greater than the wavelength λ (for $R_1 = 20$mm the distance between the apertures of Vivaldi antenna 40mm, $= 30$mm) leads to the appearance of diffraction lobes.

The calculations showed that the expansion of wedge's edges upward (like a horn) leads to a slight increase of gain.

Figures 9-12 show characteristics of antenna A_1 depending on value of radius $R_2 = 5$mm (curve——), 10mm (curve – – –), 15mm (curve - - - -) and 20MM (curve ....) for $R_1 = 2$mm (converging upward face of the wedge). The design of antenna with parameters $R_1 = 2$mm; R_2=10mm is shown on the inset of figure 9.

As it is seen (figures 9-11), if VSWR of this model of A_1 is not worse than VSWR of the previous one (figure 7), then efficiency and gain are slightly lower at almost all frequencies (gain on 1-2dB).

Figure 12 shows non-normalized radiation pattern of the antenna in the plane of **H**-vector at frequency of 10GHz for R_2 = 5mm (curve——), 10mm (curve – – –), 15mm (curve - - - -) and 20mm (curve ....), R_1=2mm.

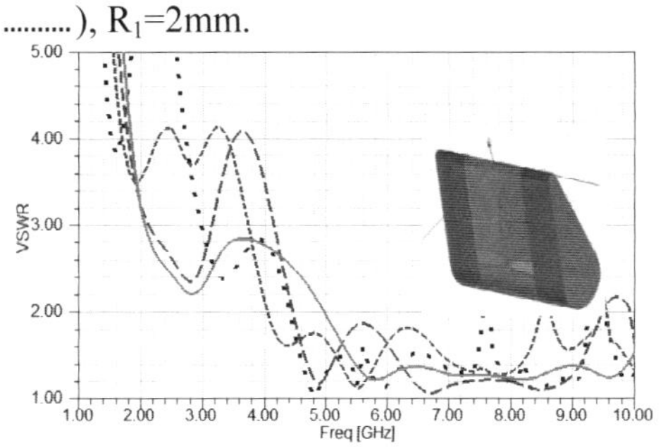

Figure 9. Dependence of VSWR on frequency of A_1 antenna for various $R_1 = 5mm; 10mm; 15mm$ and $20mm$

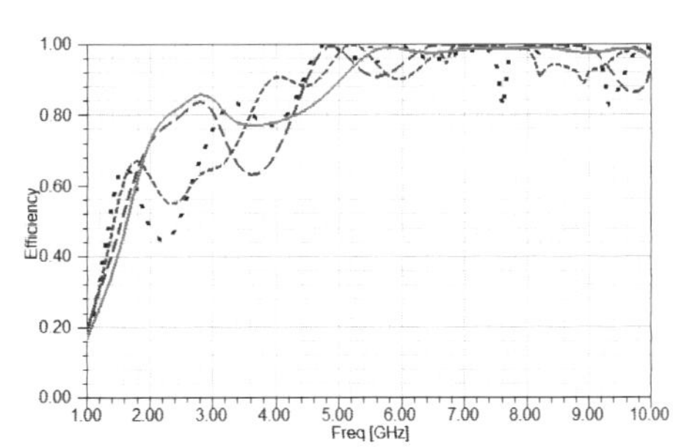

Figure 10. The dependence of the efficiency of the antenna A_1 on frequency for different $R_1 = 5mm; 10mm; 15mm$ and $20mm$

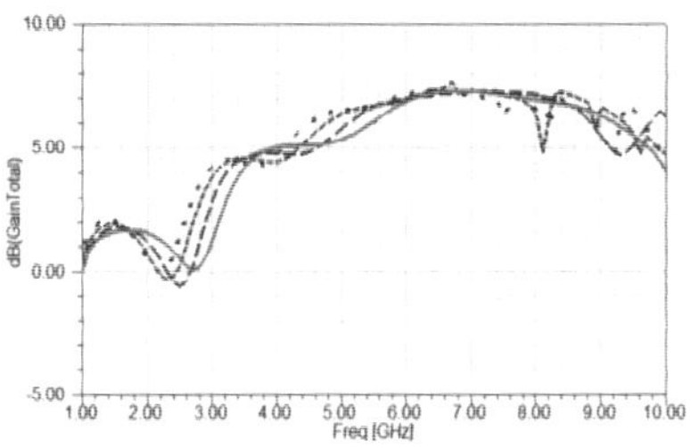

Figure 11. The dependence of the gain of antenna A_1 on frequency

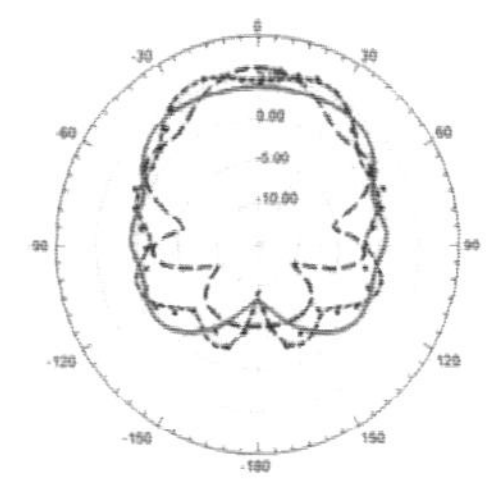

Figure 12. Radiation pattern of A_1 antenna at 10GHz

Therefore, the directional properties of the latter structure of the antenna (see inset on figure 9) are more preferable to previous, especially in wideband scanning antenna arrays.

Conclusion

The studies have shown that, changing the number of antennas and the size of the structure surface where the antennas are placed, it is possible to achieve an optimal balance between VSWR and the gain in a wide range of frequencies. Such antennas can be used as the flush-mounted ones on many types of objects with the components of constructive elements in the form of the wedge-cylinder conjunction.

Acknowledgement

This research was financially supported by Russian Science Foundation (project No. 16-19-10537)

References

[1] Baum C E and Farr E G 1993 *Impulse radiating antennas, Ultra-wideband, short-pulse electromagnetics*, Plenum Press.

[2] Schantz H 2005 *The art and the science of ultrawideband antennas* (Artech House, Inc., USA).

[3] Shin J and Schaubert D 1999 *A parameter Study of Stripline-fed Vivaldi Notch-antennas Arrays* Trans. Antennas and Pro pag., vol. 47, no5, , p.879-886.

[4] Janaswamy R and Schaubert D 1987 *Analysis of Tapered Slot Antenna* IEEE Trans. Antennas and Propag., vol. 35, no.9, pp. 1058-1065.

[5] Yukhanov Y V, Bolov R B and Privalova T Y 2010 *Arrayed Vivaldi radiators characteristics* Ultrawideband and Ultrashort Impulse Signals, Ultrawideband and Ultrashort Impulse Signals (UWBUSIS), 2010 5th International Conference on. DOI: 10.1109/UWBUSIS.2010.5609128

[6] Yukhanov Y V, Semenikhin A I, Privalova T Y and D.V. Semenikhina 2012 *Wideband radiators of scanning antenna array* Antenna Technology and Applied Electromagnetics (ANTEM), 2012 15th International Symposium on. Pages: 1 - 5, DOI: 10.1109/ANTEM.2012.6262339

IOP Conf. Series: Materials Science and Engineering **198** (2017) 012012 doi:10.1088/1757-899X/198/1/012012

Radio Frequency Interference Detection using Machine Learning.

Olorato Mosiane[1], **Nadeem Oozeer**[1,2,3], **Arun Aniyan**[1,5] **and Bruce A. Bassett**[2,3,4]

[1]SKA South Africa, The Park, Park Road, Pinelands, Cape Town 7405, South Africa.
[2]African Institute for Mathematical Sciences, 6-8 Melrose Road, Muizenberg 7945, South Africa.
[3]Centre for Space Research, North-West University, Potchefstroom, 2520, South Africa.
[4]Department of Mathematics and Applied Mathematics, University of Cape Town, 7701, Cape Town, South Africa.
[5]Department of Physics and Electronics, Rhodes University, 6139, Grahamstown, South Africa

E-mail: nadeem@ska.ac.za

Abstract. Radio frequency interference (RFI) has plagued radio astronomy which potentially might be as bad or worse by the time the Square Kilometre Array (SKA) comes up. RFI can be either internal (generated by instruments) or external that originates from intentional or unintentional radio emission generated by man. With the huge amount of data that will be available with up coming radio telescopes, an automated aproach will be required to detect RFI. In this paper to try automate this process we present the result of applying machine learning techniques to cross match RFI from the Karoo Array Telescope (KAT-7) data. We found that not all the features selected to characterise RFI are always important. We further investigated 3 machine learning techniques and conclude that the Random forest classifier performs with a 98% Area Under Curve and 91% recall in detecting RFI.

1. Introduction

Any radio signal other than the desired astronomical signal is called an unwanted signal, or spurious radiation and classified as Radio Frequency Interference (RFI). [1] & [2] summarised RFI as having a complex nature. The Square Kilometre Array (SKA) is expected to have 100-fold sensitivity and collecting area compared to known radio telescope. Apart from the RFI pick-ups through the primary beam, the peaks of the far sidelobes will tend to be picking up more unwanted signals. The establishment of a consolidated database for the SKA South Africa observatory is very crucial for various science application. RFI data resulting from the many campaigns since 2005 to the present resides currently in an assortment of small databases and test reports. As part of this project, the integration and validation of already measured RF signals in and around the core is essential.

In this paper we investigate the application of Machine learning techniques to detect RFI from data collected using the Karoo Radio Telescope (KAT-7) [3]. The paper is divided as follows: Section(2) describes the methodology adopted in pre-processing the data. Section(3) discusses the Machine Learning techniques and feature selection process together with the implication on the RFI detection. The analysis and results are discussed in Section(4) and we finally conclude the paper in Section(5).

2. Methodology: Outlier detection

Outliers in observations are data points that do not follow the statistical distribution of the overall data under investigations. There are various ways that outliers are generated, for example noisy system, human related errors, etc. In order to be able to make sense of the data, pre-processing is required.
The data used in this study were taken from one the observations done by the KAT-7. The datasets have been flagged by AO Flagger software [4], which is semi automated and validated by radio astronomers.
Communication signals are intrinsically 100% polarized and [5] describes an example how to use the polarized and unpolarized intensity to detect RFIs. Once the data have been obtained, all the Stokes parameters (I, Q, U, V) are extracted. These are taken as additional layers which represent the original data giving more dimensions from which features can be extracted. From each layer outliers are removed using the modified Z-score, which is given by the equation below:

$$M_i = \frac{0.6745(\mathrm{x}_i - \tilde{\mathrm{x}})}{MAD} \tag{1}$$

Where the Median Absolute Deviation, $MAD = \mathrm{median}(\|\mathrm{x}_i - \tilde{\mathrm{x}}\|)$ with $\tilde{\mathrm{x}}$ as the median nd x_i are the data points.
Using the modified Z-score anything larger than 3.5 is considered as outlier [6]. The outlier removed data are used to normalize the layer. The normalized layer is then unfolded in such a way that all frequency time series are taken and stitched next to each other to form a larger 1-D pseudo-time series. A similar approach is taken for the spectrum. That is for each time a spectrum is taken and all spectrum are then stitched next to each other to form a larger "spectrum". With this new perspective of the data one can now visually see periodic RFI as well as wideband RFI features. The normalization of the modified Z-score filtered, layers allows the low level RFI to shoot up.
Downgrading or smoothing the astronomical signal is very sensitive and therefore, filtering seem more appropriate to use in our case. However, filtering always goes at the expense of changing the data structure. We therefore apply the filtering process after the outlier detection procedure. Once the data have been filtered using a Gaussian, statistical analysis is applied to extract features that are fed into a machine learning algorithm.

3. Machine Learning

With the advent of new advanced technologies and the increasingly innovative capacity to gather a massive amount and variety of information, the world is faced with a deluge of data. This is already happening in the radio astronomy community. The MeerKAT data size is expected to be $\sim$ 50 Terabyte for an 1 hour observing run. However, post-processing conventional RFI detection and excision need to be automated. The algorithm presented here serve as a test case to optimize the flagging procedures already in place.

3.1. Data

The data used in this study were taken from one the observations were done by the Karoo Array Telescope (KAT-7)[3]. The KAT-7 is an array of 7, 12m diameter radio telescopes operating between 1.2 GHz-1.9 GHz. It is a precursor technology to the upcoming MeerKAT. The KAT-7 datasets we investigated were observed at a sampling rate of 0.5 seconds with 644 and 1024 channels. The number of baselines was 15 and 21 respectively. The datasets have been flagged by AO Flagger software [4], which is semi-automated and validated by radio astronomers.

3.2. Feature Extraction

Machine learning algorithms by default are not designed to take raw data in the form of images or signals as learning inputs. Therefore, the data has to be transformed into a different form, which has representative signatures of the input raw data. Feature extraction is the process of transforming the raw data into a different form such that it holds signature properties of the input class.

In this work we took 8 second segment sliding window of the signals and extracted different statistical features. The features are extracted from the spectral and time series information and are as follows: The mean of the entire channel and the kurtosis, skewness, the maximum of the cumulative sum (cumsumx), the varience (var) ,the percentile 25 and 75 (per25 and per75) and lastly the sum of the 8 second segment (sumvalues).

3.3. Machine Learning Techniques

At first three machine learning algorithms are used to classify the data. The Three algorithms are K-Nearest Neighbour (k-NN), Random Forest Classifier (RFC), Naive Bayesian (NB). In our experiment a random 70/30 % split of the data is done first. That is 30% of the data were withheld for testing the classifier and Cross-validation was performed with 70% of the data to train the algorithm. This split is essential in the true test of comparing classifiers so that the test is not biased [7].

3.4. Receiver Operating Characteristic Curve and Area Under the Curve

The Receiver Operating Characteristic Curve (ROC) is a tool used to evaluate a classifier [8]. The ROC curve is a graphical representation of the true positive rate versus the false positive rate of a binary classifier as the threshold varies between [0,1] in 0.01 increments as in [9].

The ROC curve is a tool used to study the trade-off between the false positives and the false negatives. ROC curve is a plot of True Positive Rate versus the False Positive Rate at different threshold values [9]. From the ROC curve by taking the area under the curve (AUC) one will obtain the probability that a classifier will correctly classify a randomly chosen instance [7]

While investigating all the above ML techniques, we realized that not all the features might be useful and need to be investigated [10]. There are 3 general classes of features selection algorithms: filter methods, wrapper methods and embedded methods. In this paper we will focus mainly on the wrapper methods. These types of methods take the feature importance measure obtained by a classification method and they come at a computational expense, but the result is a view of the importance of each feature in the feature space.

4. Results and Discussions

Our data contain much more non-RFI data than RFI data. This may cause a bias towards the detecting false positives. The training set was hence split up randomly 15 times in order to take into account the above mentioned bias.

The best performing classifier is the RFC and this is true for all the other splits. Since the RFC performs well in all the data splits it suggests that there is a feature or set of features that ranks higher than the other features. Because the RFC is basically a set of small decision trees it means the there is a tree or set of trees that are given more weighting to them than the others, this allows one to calculate which features are important and the features' importance. The Features that are highly ranked are those that have been extracted along each time series. This suggests that time domain information seems to be more important than spectral information in detecting RFI.

The Top 5 features were taken and the procedure of splitting and evaluation were repeated. The Best performance split is finally summarized in Table 1

Table 1. Best performing scores of the best data split with top 5 features.

ML Tech	Type(Data/RFI)	Scores			
		AUC	precision	Recall	F1-score
NB	Data	0.80	0.99	0.95	0.97
	RFI		0.07	0.32	0.12
k-NN	Data	0.86	1.00	1.00	1.00
	RFI		0.84	0.54	0.66
RFC	Data	0.98	1.00	1.00	1.00
	RFI		0.96	0.91	0.93

From Table 1 the RFC and k-NN have an overall improvement whereas the NB has has performed slightly worse but with these 5 features the best performing machine learning technique is the Random Forest Classifier.

5. Conclusion

With the deluge of astronomical data coming from new telescopes such as MeerKAT, conventional RFI detection and excision will need to be re-evaluated. We have shown that machine learning techniques can be applied to detect RFI. Pre-processing and understanding the data is a very time consuming process. However, a lot of care should be taken in pre-processing the data. From our preliminary analysis we have shown that the random forest classifier seems to produce positive results. Further investigation needs to be carried out in order to get a more robust algorithm by using a bigger sample of flagged and un-flagged raw visibility data.

Acknowledgment

The authors would like to acknowledge assistance, support and encouragement from the SKA Africa colleagues and friends who have contributed to this work.

References

[1] Pandey V and Shankar N U 2005 *URSI General Assembly Proceedings* URL `http://www.ursi.org/Proceedings/ProcGA05/pdf/J03b.5(01214).pdf`
[2] Fridman P and Baan W 2001 *Astronomy & Astrophysics* **378** 327–344
[3] Foley A, Alberts T, Armstrong R, Barta A, Bauermeister E, Bester H, Blose S, Booth R, Botha D, Buchner S *et al.* 2016 *Monthly Notices of the Royal Astronomical Society* **460** 1664–1679
[4] Offringa A 2010 *Astrophysics Source Code Library* **1** 10017
[5] Deshpande A A 2005 *Radio science* **40**
[6] Seo S 2006 *A review and comparison of methods for detecting outliers in univariate data sets* Ph.D. thesis University of Pittsburgh
[7] Buisson L d, Sivanandam N, Bassett B and Smith M 2014
[8] Kim E J, Brunner R J and Kind M C 2015
[9] McCauliff S D, Jenkins J M, Catanzarite J, Burke C J, Coughlin J L, Twicken J D, Tenenbaum P, Seader S, Li J and Cote M 2015 *The Astrophysical Journal* **806** 6
[10] Guyon I and Elisseeff A 2003 *The Journal of Machine Learning Research* **3** 1157–1182

IOP Conf. Series: Materials Science and Engineering **198** (2017) 012013 doi:10.1088/1757-899X/198/1/012013

Classifying bent radio galaxies from a mixture of point-like/extended images with Machine Learning.

David Bastien[1], **Nadeem Oozeer**[2,3,4], **Radhakrishna Somanah**[1,5]

[1] Physics Department, University of Mauritius, Reduit, Mauritius.
[2] SKA South Africa, The Park, Park Road, Pinelands, Cape Town 7405, South Africa.
[3] African Institute for Mathematical Sciences, 6-8 Melrose Road, Muizenberg 7945, South Africa.
[4] Centre for Space Research, North-West University, Potchefstroom 2520, South Africa.
[5] Universite des Mascareignes, Avenue de La Concorde, Roche Brunes, Rose Hill, Mauritius.

E-mail: `nadeem@ska.ac.za`

Abstract.
The hypothesis that bent radio sources are supposed to be found in rich, massive galaxy clusters and the avalibility of huge amount of data from radio surveys have fueled our motivation to use Machine Learning (ML) to identify bent radio sources and as such use them as tracers for galaxy clusters. The shapelet analysis allowed us to decompose radio images into 256 features that could be fed into the ML algorithm. Additionally, ideas from the field of neuro-psychology helped us to consider training the machine to identify bent galaxies at different orientations. From our analysis, we found that the Random Forest algorithm was the most effective with an accuracy rate of 92% for a classification of point and extended sources as well as an accuracy of 80% for bent and unbent classification.

1. Introduction

Galaxies can be observed at different wavelengths, each of these wavelengths reveals different morphology and characteristics of galaxies. Galaxies hosting an active galactic nuclei (AGN) are usually powerful radio sources that produce jets and extended radio emitting regions known as lobes. These sources also show various radio morphologies, from single point like source to extended ones. The lobes are huge and can range from 100 pc up to few Mpc in linear extent and can be bent due to different mechanisms.

The first step for source classification is to do a source finding. This will basically fit Gaussian at various local maxima and extract various fitting parameters such as the source Gaussian parameters(the source major and source minor axes) and also the position angle. The sources will then be classified as point/un-resolved or resolved/extended depending by how much the fitted parameters to the sources deviate from the synthesized beam parameters of the telescope. In order to check whether sources are connected or not, contours are usually drawn overlay onto Optical/Infra Red (IR) images. The core of the radio source (if present) normally align with the optical/IR counterpart. With the upcoming deep surveys, the volume of data/images that will be produced will be huge. Manual classification will be a tedious if not impossible. A Machine Learning (ML) approach can be a vital tool for classification analysis.

IOP Conf. Series: Materials Science and Engineering **198** (2017) 012013 doi:10.1088/1757-899X/198/1/012013

The goal of the work presented here shows that ML techniques are efficient and robust. ML can be put to use to analyse large amount of data that will result from future telescopes like the Square Kilometre Array (SKA) and its pathfinders such as MeerKAT and ASKAP. This paper is divided as follows: §2 describes Machine Learning techniques and feature selection process adopted in in this work. The results and analysis are discussed in §3 and we finally conclude the paper in §4.

2. Machine Learning application

ML algorithms can be broken down into two main categories: (i) Supervised and (ii) Unsupervised learning. Each one of them can be further broken down into classification and regression for supervised learning as well as clustering and dimensionality reduction for unsupervised learning.

In order to analyse the radio images a sample was made from various already classified radio sources. All the images were downloaded into small postage images from the Skyview Virtual Observatory [1]. The first step was to carry out the point-like/extended classification and second step was the straight/bent classification. Before applying any ML algorithm, features characterising the radio sources were extracted. For example [1] made use of the parameters of the fitted Gaussians. In our case coefficients obtained from the shapelet analysis were used. Figure (1) shows the shapelet decomposition for a typical image.

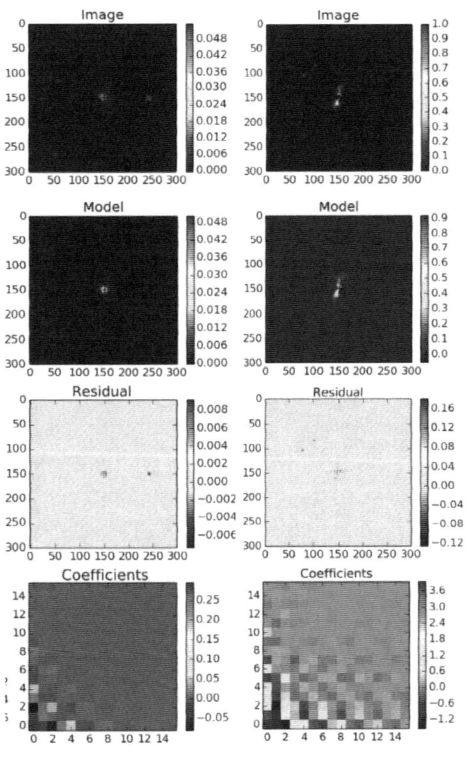

Figure 1. The decomposition of 2 radio sources NGC 0383 (1st column) and NGC 1365 (2nd column). From top to bottom we have the radio Image, the model (the part of the image that has been modeled by shapelets), the residual (part not modeled in the coefficients), The shapelet oefficients (16×16 matrix). Shapelets decomposition can be defined as a linear decomposition of an object into a series of localized 1D basis functions with different shapes [2]. The localized basis functions will be orthonormal with each other and are a set of weighted polynomial given by:

$$\phi_n(x) = \left[2^n \pi^{\frac{1}{2}} n!\right]^{-\frac{1}{2}} H_n(x) e^{-\frac{x^2}{2}}$$

Where n is a non-negative integer and $H_n(x)$ is the Hermite polynomial. For this project 2D Shapelets were used. The basis 2D functions can be obtained by taking the product of two 1D basis functions.

The first 256 shapelet coefficients were extracted for each images and arrange as vectors. Each radio source is a vector in this vector space. The direction of the vector gives the shape and the orientation while its amplitude is a function of the total flux.

[1] http://skyview.gsfc.nasa.gov

Table 1. Results Machine Learning for point-like and extended sources classification. The confusion matrix shows the number of sources and how they were classified by the algorithms. For example, in the first row 65 sources were classified correctly as point while 15 point sources were classified as extended.

Machine Learning Techniques	AUC	Accuracy	Recall	Precision	F1-Score	Confusion Matrix		
							Point	**Extended**
Random Forest (RF)	0.94	0.92	0.67	0.81	0.73	**Point**	65	32
						Extended	15	460
							Point	**Extended**
k-Nearest Neighbour (k-NN)	0.94	0.90	0.77	0.67	0.72	**Point**	75	22
						Extended	37	438
							Point	**Extended**
Adaboost (Ada)	0.94	0.91	0.73	0.75	0.74	**Point**	71	26
						Extended	24	451
							Point	**Extended**
Naive Bayes (NB)	0.89	0.85	0.89	0.53	0.67	**Point**	86	11
						Extended	75	400

Once normalised only the information about the shape is stored as a vector. The comparison of the sources can be made by finding the distances between the various vectors. If two sources have the exact same shape and orientation with different peak amplitude once the shapelet vector coefficients normalised, their projected vector will be equal. This technique is useful since it enabled us to compare faint objects with very bright objects with similar shapes.

Before inputting our coefficients into the ML algorithm, a visualisation is carried out in order to give us an idea of how well features are and to what extent our point and extended sources get separated in the feature space. A dimensionality reduction is required since we are dealing with 256 dimensions. The Isomap [3] also referred to as Isometric Mapping was used. It is an algorithm that maps manifold on our high dimensional data and projects it on lower dimensions, it was implemented using the ***sci-kit*** [4] learn module in Python. It takes as input the set of vectors that have been described and the vectors are reduced to 3-D and further plotted in 2-D.

Our result for the isomapping (plotted on a x-y grid ranging from [-1,1]) shows that most of the data points lie in the region $-0.5 \leq y \leq 0$ and $-0.5 \leq x \leq +0.5$. By inspecting the radio images we found that the radio galaxies in this region appears frequently in the data set and they are single Gaussian. A centre point can be drawn at $x = 0$ and $y = -0.25$ and data points away from this centre corresponds to images for which more than 1 Gaussian can be fitted. Those centred at $x = 0$, $y = -0.25$ are can be fitted with Gaussians that are circular.

3. Results and Discussions

Once convinced of the features selected from the images from [5], we input the whole 256 coefficient into the ML algorithm. Four algorithms have been used; k-Nearest Neighbour (k-NN), Adaboost, Random Forest (RF) and Naïve Bayes (NB).

We make use of a stratified cross-validation techniques and with an Receiver Operating Characteristic (ROC) plot we could visualise the result of each techniques. Three of the algorithms had the same ROC Area under curve (AUC) of 0.94. In such a case other performance metrics are used as an efficiency indicator. The RF has the best accuracy and precision but NB had the best Recall and Adaboost had the best F1-Score, table(1).

The dataset from [6], was chosen as it provides a sample of classified Bent and Unbent sources at high resolution. The Bent-Unbent classification requires and additional rotational training as compared to the point extended classification. This is a new feature extraction process where the machine is trained with sources that have been rotated at different angles. In this new training scheme the source is rotated from $0°$ to $360°$ in steps of $10°$. At each step, the source is decomposed and the shapelets coefficient are stored.

To ensure that the same source at different orientation was not used in training and testing, the following procedures were adopted to cross-validate the results:

- The number of folds n is a multiple of 369. In that case n=9 is the best fold number.
- An unstratified cross-validation is used, this ensures that the records are not scrambled and hence that the same source does not appear in different folds.

For the case of Bent-Unbent classification from extended sources, a similar approach as explained above and that used by [7] was applied. 5 metrics based on the true positive(t_p), true negative (t_n), false positive (f_p), false negative (f_n) were used; the accuracy, the recall, the precision, the F1-score and the ROC AUC curve, to define the efficiency the algorithms. The RF was the best classifier for bent and unbent sources, with an AUC of 0.9, 80% Accuracy, 93% Recall and 86 % F1-score. The k-NN had the largest precision of 84% in this case.

From our analysis, the Naive Bayes is here the black sheep of the family, the latter has an ROC which has an AUC less 0.5. This shows that the Naive Bayes classifier is worst that a random classifier and cannot be used in our analysis.

4. Conclusion

We showed that using shapelets coefficients as features from radio images and Random Forest as a classifier yield an accuracy of 92% in classifiying point-like from extended sources. Through the use of vectors and visualisation tools like Isomap and t-distributed stochastic neighbour embedding (t-SNE) a geometrical approach was brought forward to implement the classification problem. A rotational training technique inspired from a neuro-psychological phenomena known as mental rotation was implemented. The ML algorithms were then trained using sources at different orientations. We made use of a dataset of bent and unbent sources and using Random Forest we achieved an accuracy of 80% in properly classifying them.

5. Acknowledgement

This research has been conducted using resources provided by the Science and Technology Facilities Council (STFC) through the Newton Fund and the SKA Africa.

References

[1] D. D. Proctor. Comparing Pattern Recognition Feature Sets for Sorting Triples in the FIRST Database. *ApJS*, 165:95–107, July 2006.

[2] A. Refregier. Shapelets - I. A method for image analysis. *MNRAS*, 338:35–47, January 2003.

[3] Joshua B. Tenenbaum, Vin de Silva, and John C. Langford. A global geometric framework for nonlinear dimensionality reduction. *Science*, 290(5500):2319, 2000.

[4] F. Pedregosa, G. Varoquaux, A. Gramfort, V. Michel, B. Thirion, O. Grisel, M. Blondel, P. Prettenhofer, R. Weiss, V. Dubourg, J. Vanderplas, A. Passos, D. Cournapeau, M. Brucher, M. Perrot, and E. Duchesnay. Scikit-learn: Machine learning in Python. *Journal of Machine Learning Research*, 12:2825–2830, 2011.

[5] S. van Velzen, H. Falcke, P. Schellart, N. Nierstenhöfer, and K.-H. Kampert. Radio galaxies of the local universe. All-sky catalog, luminosity functions, and clustering. *A&A*, 544:A18, August 2012.

[6] J. D. Wing and E. L. Blanton. Galaxy Cluster Environments of Radio Sources. *AJ*, 141:88, March 2011.

[7] L. du Buisson, N. Sivanandam, B. A. Bassett, and M. Smith. Machine learning classification of SDSS transient survey images. *MNRAS*, 454:2026–2038, December 2015.

IOP Conf. Series: Materials Science and Engineering **198** (2017) 012014 doi:10.1088/1757-899X/198/1/012014

MIGHTEE: The MeerKAT International GHz Tiered Extragalactic Exploration

A Russ Taylor[1,2] and Matt Jarvis[2,3]

[1] Inter-University Institute for Data Intensive Astronomy, and Department of Astronomy, University of Cape Town, Rondebosch 7701, South Africa

[2] Department of Physics and Astronomy, University of the Western Cape, Bellville 7537, South Africa

[3] Department of Physics, University of Oxford, Oxford OX1 3RH,UK

E-mail: russ@ast.uct.ac.za

Abstract. The MeerKAT telescope is the precursor of the Square Kilometre Array mid-frequency dish array to be deployed later this decade on the African continent. MIGHTEE is one of the MeerKAT large survey projects designed to pathfind SKA key science in cosmology and galaxy evolution. Through a tiered radio continuum deep imaging project including several fields totaling 20 square degrees to microJy sensitivities and an ultra-deep image of a single 1 square degree field of view, MIGHTEE will explore dark matter and large scale structure, the evolution of galaxies, including AGN activity and star formation as a function of cosmic time and environment, the emergence and evolution of magnetic fields in galaxies, and the magnetic counter part to large scale structure of the universe.

1. Introduction

The MeerKAT radio telescope is nearing completion on the Karoo desert in South Africa. MeerKAT is a precursor of the Square Kilometre Array mid-frequency dish array (SKA1-mid), and following several years of operation as a South African telescope, it will be incorporated in to the SKA phase 1 facility. The first 32 elements of MeerKAT will be release for shared-risk science programs in 2017, and the full array is expected to be operational by early 2018.

The MeerKAT science program will consist primarily of key science, legacy-style Large Survey Projects (LSPs). The LSPs are direct pathfinders of key science projects and observations being planned for the SKA mid-frequency array. The MeerKAT International GHz Tiered Extragalactic Exploration project (MIGHTEE) is one of the approved MeerKAT LSPs, with the aim explore the high-redshift radio universe with very deep, broad-band, wide-field, radio continuum and spectral line imaging. This paper briefly describes the MIGHTEE observing plan and science goals.

2. Observing Plan

2.1. MeerKAT

MeerKAT is 64-element array of 13.5-m parabolic antennas laid out in a centrally concentrated configuration with a maximum baseline of about 8 km. Fourty-eight of the antennas are situation in a central 1-km diameter core. The antennas have offset Gregorian optics, providing an

IOP Conf. Series: Materials Science and Engineering **198** (2017) 012014 doi:10.1088/1757-899X/198/1/012014

Table 1. MeerKAT Imaging Specifications

Frequency Band	Resolution (arcseconds)	Field of View (degrees)	RMS noise in one hour (microJy)
1.75 – 2.50 GHz	4	0.65	~5
0.90 - 1.67 GHz	6	1.1	4.7
0.58 - 1.01 GHz	10	1.75	9.0

unobstructed radiation path to the primary reflector and very high forward gain, low-sidelobe levels and excellent aperture efficiency.

The relatively small diameter of the MeerKAT antennas coupled with the large number of antennas and large total collecting makes it a powerful wide-field imaging instrument, with a combination of large field-of-view, wide-bandwidth, high sensitivity and outstanding sampling of the Fourier transform of the sky with 2016 instantaneous baseline samples. Imaging specifications of MeerKAT for the frequency ranges that will be used for the MIGHTEE project are listed in Table 1. At the lowest frequency band the bandwidth is limited by the available RF to 440 MHz. At the mid- and high-frequency bands the digitizer captures an instantaneous bandwidth of 770 MHz. The correlator provides four correlation products, allowing imaging in all polarization products.

2.2. Observations

MIGHTEE will use the imaging capabilities of MeerKAT to carry out a tiered exploration of the extragalactic sky via a tier-1 medium deep survey to an rms noise level of about 1 microJy over 20 deg^2, and a tier-2 ultra-deep survey to about 200 nanoJy over 1 deg^2. With up to 32,768 channels, each 3.27 kHz wide, MIGHTEE data cubes will image both the broad-band spectro-polarimetric properties of continuum emission from galaxies and the narrow-band emission from the spectral line of atomic hydrogen.

The tier-2 survey will be observed commensally with a deep HI spectral line survey (LADUMA). It will comprise a single pointing of the Chandra Deep Field South (CDFS) covering the FoV of the telescope at the two lowest bands. For a planned total of 330 hours of integration at the mid-band (0.9 – 1.67 GHz) MIGHTEE will achieve a theoretical rms noise of 260 nanoJy. At the low band (0.58 – 1.0 GHz) 3090 hours of integration will provide a full-band image with rms noise of 160 nanoJy.

The 20 deg^2 medium deep tier-1 survey has a planned total observing time of 1920 hours distributed over four fields: the XMM-LSS (6.7 deg^2), E-CDFS (8.3 deg^2), ELAIS-S1 (1.6 deg^2) and the COSMOS field (1 deg^2). XMM-LSS, E-CDFS and ELAIS-S1 will be covered by mosaicking of multiple pointings as illustrated in Fig 1. XMM-LSS will be observed at the mid-frequency band only. We will cover 4 deg^2 of E-CDFS and the COSMOS field with both the mid- and high- bands, providing increased sensitivity and resolution and total bandwidth of 0.9 – 2.5 GHz.

Multi-wavelength information is crucial to the science goals of MIGHTEE. Each of the chosen fields has extensive ancillary information from optical and infrared imaging surveys. The layout of the radio coverage of each field was chosen to maximize overlap with such surveys. Rectangular and circular regions in Fig. 1 indicate coverage by the VEILS (pink), VIDEO (blue) and DES (white) image sets.

IOP Conf. Series: Materials Science and Engineering **198** (2017) 012014 doi:10.1088/1757-899X/198/1/012014

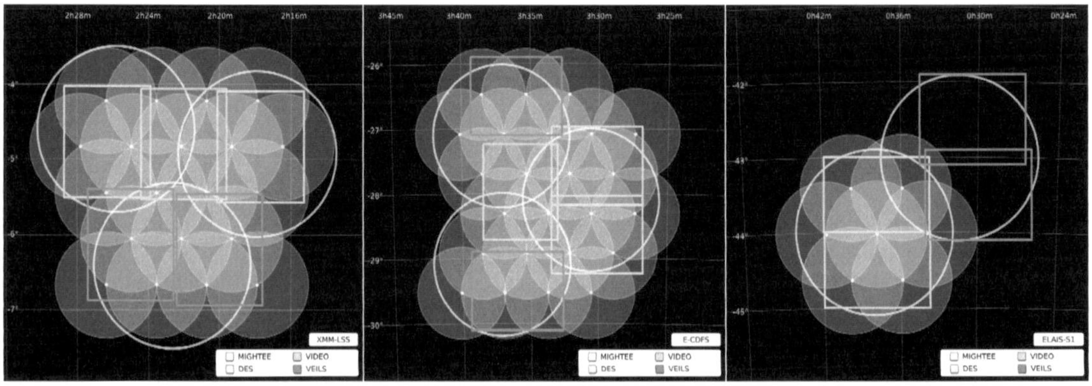

Figure 1. Pointing strategies for the three mosaic imaged MIGHTEE fields. Left: XMM-LSS (20 pointings, 6.7 deg^2). Middle: E-CDFS (24 pointings, 8.3 deg^2). Right: ELAIS-S1 (7 pointings, 1.6 deg^2). The yellow shaded circles indicate the pointing locations and primary beam areas of the MeerKAT observations. Figure compliments of Ian Heywood.

3. MIGHTEE Science

In concert with multi-wavelength data the MIGHTEE radio spectro-polarimetric continuum and spectral-line image cubes are designed to provide a unique and powerful data set to address scientific questions in large-scale structure and cosmology, galaxy evolution, and cosmic magnetism.

3.1. Large-Scale Structure and Cosmology

The depth and bread of the MIGHTEE and will also allow us to measure the cosmological bias, e.g. how radio sources of different types (e.g. star forming galaxies, radio galaxies, AGN) trace the underlying dark matter distribution. It is crucial to do this in fields with the best multi-wavelength data as it allows us to accurately disentangle the different types of sources, whilst obtaining the required number density (at least of all types of source apart from FRIIs, which are intrinsically rare).

MIGHTEE will yield a grid of Rotation Measures (RM) toward background sources with sufficient area density and precision to detect and measure the properties of magnetic fields embedded in the large-scale structure - the magnetic cosmic web. Simulations predict that the IGMF in filaments would induce excess Faraday Rotation Measures with a flat second-order structure function (SF) of $\sim$100 rad^2 m^{-4} for angular separation of $r > 0.1$ deg. The power from this contribution to the RM structure function will be distinguishable from the foreground of the Milky Way galaxy on angular scales smaller than a few degrees [1]. MIGHTEE will probe this structure function with the the required number density and precision, and over angular scales of a few arcminutes to several degrees, offering our best opportunity before the SKA to use this technique to detect the magnetic cosmic web.

The coherent distortion of distant galaxy shapes due to the gravitational effect of dark matter (weak lensing) is one of the most powerful probes of dark matter and dark energy. Radio polarization position angle provide unique information on the intrinsic (unlensed) shape of background galaxies that is unaffected by lensing. This effect can be used to mitigate the primary error in weak lensing experiments.

3.2. *Galaxy Evolution*

While the stellar properties of galaxies spanning the full range of masses over a large range of redshifts have been well studied, our knowledge of the neutral gas (HI) content of these same galaxies is restricted to the local universe. If we hope to understand the build-up of stellar mass, we must observe the neutral gas reservoir of fuel from which the molecular gas forms, eventually turning into stars, along with the interface between this gas and the galaxies and environment in which it is located. MeerKAT will open up new parameter space in the study of the neutral gas component of galaxies.

Galaxies follow known scaling relations, such as the so-called star-formation main sequence [2; 3] relating star formation with stellar mass. Underlying these relations is a complex cycle of acquisition, storage, consumption, expulsion and re-acquisition of gas, acting to regulate a galaxy's ability to form stars [e.g. 4]. With MIGHTEE we will not only trace the origins of the gas that eventually turns into stars, but with the extreme sensitivity to radio-continuum emission, we will also measure the end-point, namely the star-formation rate. MIGHTEE has the sensitivity and frequency coverage to detect a statistically significant number of galaxies in atomic hydrogen out to $z > 0.2$, and to measure star formation rates from continuum emission to $z > 3$. Together with optical and near-IR data we will trace the star formation history and the evolution of the ISM in galaxies over a significant fraction of cosmic time and diverse environments.

MIGHTEE will detect virtually every active radio galaxy to the edge of the universe in each field. The broad-band polarization data will allow new approaches to measuring the intrinsic properties of AGN core and jet environments.

3.3. *Magnetic field in Galaxies and Clusters*

Through its ability to detect polarization of sources at high redshift, MIGHTEE will be a cornerstone for investigation of the evolution of cosmic magnetic fields. For disk galaxies, this range is $z < 2.5$, and for AGN and starbursts it reaches out to $z = 7$, into the epoch of reionization. From the integrated polarization properties of galaxies we will trace the emergence and growth of coherent magnetic field in galaxies and in AGN and its relation to star formation.

It is now well established that the intracluster medium (ICM) of galaxy clusters, which is composed mainly of thermal gas emitting in the X-ray energy band, is permeated by magnetic fields over the entire cluster volume [see 5, for a review]. The MIGHTEE RM grid will enable us to probe the spatial structure function of magnetic fields within galaxy clusters, offering the best opportunity before the SKA to constrain the redshift evolution of cluster magnetic fields and providing the first opportunity to test early seed field models for the origin of cosmic magnetic fields.

References

[1] Akahori T, Gaensler B M and Ryu D 2014 *Astrophysical Journal* **790** 123

[2] Noeske K G, Weiner B J, Faber S M, Papovich C, Koo D C, Somerville R S, Bundy K, Conselice C J, Newman J A, Schiminovich D, Le Floc'h E, Coil A L, Rieke G H, Lotz J M, Primack J R, Barmby P, Cooper M C, Davis M, Ellis R S, Fazio G G, Guhathakurta P, Huang J, Kassin S A, Martin D C, Phillips A C, Rich R M, Small T A, Willmer C N A and Wilson G 2007 *Astrophysical Journal Letters* **660** L43–L46

[3] Johnston R, Vaccari M, Jarvis M, Smith M, Giovannoli E, Häußler B and Prescott M 2015 *MNRAS* **453** 2540–2557

[4] Davé R, Thompson R J and Hopkins P F 2016 *ArXiv e-prints* (*Preprint* 1604.01418)

[5] Ferrari C, Govoni F, Schindler S, Bykov A M and Rephaeli Y 2008 *Space Science Reviews* **134** 93–118

Institute of Physics
Dirac House, Temple Back
Bristol BS1 6BE UK

ISSN: 1757-8981
ISBN 978-1-5108-4040-9

9 781510 840409

CURRAN ASSOCIATES INC.
proceedings
.com

International Conference on the Frontiers in Atomic, Molecular, and Optical Physics (AMO 2016)

Journal of Physics: Conference Series Volume 752

Shanghai, China
23 – 26 May 2016

IOP
Institute of Physics